Study Guide for

Regression Analysis

Statistical Modeling
of a Response Variable

Second Edition

Rudolf J. Freund
Department of Statistics
Texas A & M University

William J. Wilson
Department of Mathematics and Statistics
University of North Florida

Ping Sa
Department of Mathematics and Statistics
University of North Florida

ELSEVIER

AMSTERDAM · BOSTON · HEIDELBERG · LONDON
NEW YORK · OXFORD · PARIS · SAN DIEGO
SAN FRANCISCO · SINGAPORE · SYDNEY · TOKYO
Academic Press is an imprint of Elsevier

Acquisitions Editor: Tom Singer
Project Manager: Jeff Freeland
Marketing Manager: Linda Beattie
Cover Design: Eric DeCicco
Printer: LexisNexis

Academic Press is an imprint of Elsevier
30 Corporate Drive, Suite 400, Burlington, MA 01803, USA
525 B Street, Suite 1900, San Diego, California 92101-4495, USA
84 Theobald's Road, London WC1X 8RR, UK

This book is printed on acid-free paper.

ISBN-13: 978-0-12-372504-2
ISBN-10: 0-12-372504-6

For information on all Academic Press publications
visit our Web site at www.books.elsevier.com

Preface

This manual is intended to be used with ***Regression Analysis*** *Statistical Modeling of a Response Variable,* Second Edition. It provides complete solutions, interpretation of results, and additional comments for selected end of chapter exercises. These exercises were carefully chosen to be representative of the material covered in each chapter. All exercises have been worked in SAS, and in many cases the output has been abbreviated and/or edited for compactness and readability. Slightly different results, and output formats may occur if other software is used. These solutions may serve as guides for the rest of the exercises in the chapter.

Because many of the exercises in the text, especially in the later portions, are somewhat open ended, the solutions given in this manual should not be considered to be the only solution for any given problem. The interpretation represents the thinking of the authors, and does not mean that other interpretations cannot be correctly made. Instead, these solutions are meant to be used as models for solving the rest of the exercises.

While we have checked the solutions carefully, it is possible that errors exist. We will be most grateful to receive any errata as well as suggestions for better analyses or interpretations.

Rudolf J. Freund
William J. Wilson
Ping Sa
Diery Cissokho

Table of Contents

Chapter 1
The Analysis of Means
Solutions

3. Table 1.9 gives the results of a study of the effect of diet on the weights of laboratory rats. The data are weights in ounces of rats taken before the diet and again after the diet.

SAS Code:

The data for this exercise is in reg01p07. In order to use the linear models approach, the data has been modified. A value of "0' is assigned to the variable diet for the "before" weight, and a value of "1" is assigned to the variable diet for the "after" weight. The modified data is given below and given the name "reg01p03modify".

```
data reg01p03modify;
input rat diet weight;
cards;
1     0     14
2     0     27
3     0     19
4     0     17
5     0     19
6     0     12
7     0     15
8     0     15
9     0     21
10    0     19
1     1     16
2     1     18
3     1     17
4     1     16
5     1     16
6     1     11
7     1     15
8     1     12
9     1     21
10    1     18
;
proc glm data =reg01p03modify;
class rat diet;
model weight = rat diet;
proc means;
var weight;
proc means;
by diet;
var weight;
proc sort;
by rat;
proc means;
by rat;
var weight;
run;
```

Relevant Output:

The GLM Procedure

Dependent Variable: weight

Source	DF	Sum of Squares	Mean Square	F Value	Pr > F
Model	10	217.0000000	21.7000000	5.03	0.0116
Error	9	38.8000000	4.3111111		
Corrected Total	19	255.8000000			

R-Square	Coeff Var	Root MSE	weight Mean
0.848319	12.28593	2.076322	16.90000

Source	DF	Type III SS	Mean Square	F Value	Pr > F
rat	9	200.8000000	22.3111111	5.18	0.0112
diet	1	16.2000000	16.2000000	3.76	0.0845

Output from Proc Means:

N	Mean	Std Dev	Minimum	Maximum
20	16.9000000	3.6692176	11.0000000	27.0000000

-- diet=0 --

N	Mean	Std Dev	Minimum	Maximum
10	17.8000000	4.2635405	12.0000000	27.0000000

-- diet=1 --

N	Mean	Std Dev	Minimum	Maximum
10	16.0000000	2.9059326	11.0000000	21.0000000

-- rat=1 --

N	Mean	Std Dev	Minimum	Maximum
2	15.0000000	1.4142136	14.0000000	16.0000000

-- rat=2 --

N	Mean	Std Dev	Minimum	Maximum

| 2 | 22.5000000 | 6.3639610 | 18.0000000 | 27.0000000 |

-------------------------------------- rat=3 --------------------------------------

N	Mean	Std Dev	Minimum	Maximum
2	18.0000000	1.4142136	17.0000000	19.0000000

-------------------------------------- rat=4 --------------------------------------

N	Mean	Std Dev	Minimum	Maximum
2	16.5000000	0.7071068	16.0000000	17.0000000

-------------------------------------- rat=5 --------------------------------------

N	Mean	Std Dev	Minimum	Maximum
2	17.5000000	2.1213203	16.0000000	19.0000000

-------------------------------------- rat=6 --------------------------------------

N	Mean	Std Dev	Minimum	Maximum
2	11.5000000	0.7071068	11.0000000	12.0000000

-------------------------------------- rat=7 --------------------------------------

N	Mean	Std Dev	Minimum	Maximum
2	15.0000000	0	15.0000000	15.0000000

-------------------------------------- rat=8 --------------------------------------

N	Mean	Std Dev	Minimum	Maximum
2	13.5000000	2.1213203	12.0000000	15.0000000

-------------------------------------- rat=9 --------------------------------------

N	Mean	Std Dev	Minimum	Maximum
2	21.0000000	0	21.0000000	21.0000000

-- rat=10 --

N	Mean	Std Dev	Minimum	Maximum
2	18.5000000	0.7071068	18.0000000	19.0000000

(a) Define an appropriate linear model to explain the data

Since the interest of this study is the effect of diet on the weights of laboratory rats, an appropriate model is the weight of the rats as response variable and the rat and diet as independent variables. Both diet and rat are defined to be classification variables in this model. Diet is equal to 0 before the diet and 1 after the diet. We can then represent the model as follows:

$$y_{ij} = \mu + \alpha_i + \beta_j + \varepsilon_{ij}, \ i = 1, 2 \text{ and } j = 1, ..., 10,$$

where y_{ij} = weight of the j-th rat in the i-th level of diet.
μ = the overall mean weight
α_i = the effect of the i-th diet
β_j = the effect of the j-th rat
ε_{ij} = the random error

Estimate $(\alpha_1) = 17.8 - 16.9 = 0.9$ (mean weight of all rats before diet minus overall mean)
Estimate $(\alpha_2) = 16 - 16.9 = -0.9$ (mean weight of all rats after diet minus overall mean)
Estimate $(\beta_1) = 15 - 16.9 = -1.9$ (mean weight of subject 1 before and after diet)
Similarly,
Estimate $(\beta_2) = 5.6$
Estimate $(\beta_3) = 1.1$
Estimate $(\beta_4) = -0.4$
Estimate $(\beta_5) = 0.6$
Estimate $(\beta_6) = -5.4$
Estimate $(\beta_7) = -1.9$
Estimate $(\beta_8) = -3.4$
Estimate $(\beta_9) = 4.1$
Estimate $(\beta_{10}) = 1.6$

5

(b) Using the linear models approach with $\alpha = 0.01$, test whether the diet changed the weight of the laboratory rats.

Step1: $H_0: \alpha_1 - \alpha_2 = 0$ vs. $H_A: \alpha_1 - \alpha_2 \neq 0$

Step2: $\alpha = 0.01$

Step3: The test statistic is $F^* = \dfrac{MSE_{Hypothesis}}{MSE_{Unrestricted}}$ has an F (1-α, 1, n-2) distribution under H_0.

Step4: We will reject H_0 if the $F^* >$ F (0.99, 1, 9) = 10.6

Step5: $F^* = \dfrac{16.20}{4.31} = 3.76$, $p - value = 0.0845$

Step6: Since $F^* < 10.6$, we fail to reject H_0.

There is not sufficient evidence to support the claim that the diet changed the weight of the laboratory rats at the 0.01 level of significance.

Note: an alternate approach to working this problem is to consider it a paired experiment and use the simple differences.

SAS Code:

The data for this exercise is reg01p03.
```
data reg01p03;
input rat before after ;
diff=before-after;
cards;
1  14  16
2  27  18
3  19  17
4  17  16
5  19  16
6  12  11
7  15  15
8  15  12
9  21  21
10 19  18
proc univariate;
var diff;
run;
```

Relevant Output:

Variable: diff

Moments

N	10	Sum Weights	10
Mean	1.8	Sum Observations	18
Std Deviation	2.93636207	Variance	8.62222222
Skewness	1.69313231	Kurtosis	4.23986628
Uncorrected SS	110	Corrected SS	77.6

```
                    Coeff Variation     163.131226    Std Error Mean     0.92855922
```

The model is then: $y_i = \mu + \varepsilon_i$, $i = 1, \ldots, 10$.

Where : y_i is the i-th difference, μ is the mean of the differences, and ε_i is the error term. The value μ is estimated by the sample mean difference = 1.8.

The $SS_{hypothesis} = 110-77.6 = 32.4$
The $MSE_{unrestricted} = 8.62222222$
The F with 1 and 9 degrees of freedom = $32.4/8.62222222 = 3.76$ so the result is identical.

7. Three different laundry detergents are being tested for their ability to get clothes white. An experiment was conducted by choosing three brands of washing machines and testing each detergent in each machine. The measures used were a whiteness scale, with high values indicating more "whiteness." The results are given in Table 1.12.

SAS Code:

The data for this exercise is reg01p07. In this and subsequent exercises, only the SAS programming commands are given. The input statements and the data are given in the data sets on the cd accompanying the text.

```
proc glm;
class sol machine;
model white = sol machine;
run;

proc glm;
class sol machine;
model white = machine;
run;
```

Relevant Output:

```
                         The GLM Procedure

Dependent Variable: white
                                  Sum of
        Source              DF    Squares      Mean Square    F Value    Pr > F
        Model                4   625.7777778   156.4444444     11.09     0.0194
        Error                4    56.4444444    14.1111111
        Corrected Total      8   682.2222222

                R-Square     Coeff Var    Root MSE    white Mean
                0.917264     26.00637     3.756476    14.44444


        Source              DF    Type III SS    Mean Square    F Value    Pr > F
```

sol	2	588.2222222	294.1111111	20.84	0.0077
machine	2	37.5555556	18.7777778	1.33	0.3606

Source	DF	Sum of Squares	Mean Square	F Value	Pr > F
Model	2	37.5555556	18.7777778	0.17	0.8438
Error	6	644.6666667	107.4444444		
Corrected Total	8	682.2222222			

R-Square	Coeff Var	Root MSE	white Mean
0.055049	71.76144	10.36554	14.44444

Source	DF	Type III SS	Mean Square	F Value	Pr > F
machine	2	37.55555556	18.77777778	0.17	0.8438

(a) Define an appropriate model for this experiment. Consider the difference between washing machines a nuisance variation and not of interest to the experimenters.

Since this experiment actually has two factors, the variety of the solution and the machines, we will use the "two-factor ANOVA model" with one factor considered as a block: the factor machines. Our model is a Randomized Complete Block Design (RCBD) with solutions as treatments and machines as subjects (blocks). The model can be represented in the following form:

$$y_{ij} = \mu + \alpha_i + \beta_j + \varepsilon_{ij}, \quad i = 1, 2, 3 \text{ and } j = 1, 2, 3$$

where:

y_{ij} = the response from the ith treatment (solution) and the jth block (machine).

μ = the overall mean

α_i = the effect of the ith treatment

β_j = the effect of the jth block.

ε_{ij} = the random error term

(b) Find the $SSE_{unrestricted}$ and the $SSE_{restricted}$ model.

The sum of squares error (SSE) restricted and unrestricted can be read directly from the Analysis of Variance output tables of the regression procedure of the two models. The sum of squared errors of the full model ($SSE_{unrestricted}$) is equal to 56.44. The sum of squared errors of the reduced model ($SSE_{restricted}$) is equal to 644.67.

(c) Test the hypothesis that there is no difference between detergents

Step1: $H_0 : \alpha_i = 0$ for all i vs. $H_A : \alpha_i \neq 0$ for $i = 1$ or $i = 2$ or $i = 3$

Step2: $\alpha = 0.05$

Step3: The test statistic is $F^* = \dfrac{MSE_{Hypothesis}}{MSE_{Unrestricted}} = \dfrac{(SSE_{Restricted} - SSE_{Unrestricted})/2}{MSE_{Unrestricted}}$ has an

F ($1-\alpha$, 2, 4) distribution under H_0.

Step4: We will reject H_0 if the $F^* > F$ (0.95, 2, 4) = 6.94

Step5: $F^* = \dfrac{(644.67 - 56.44)/2}{14.11} = 20.84$, $p-value = 0.0077$

Step6: Since $F^* > 6.94$, we reject H_0.

There is enough evidence to support the conclusion that at least one of the detergents is different from the others. Also, we can conclude that, using the same washing machine, the whiteness level of the cloth significantly varies according to the brand of detergent used.

Chapter 2
Simple Linear Regression
Solutions

1. Calculator Exercise

SAS Code:

The data for this exercise is reg02p01.

```
proc reg data = reg02p01;
model y = x /p cli clm/*predicted values, confidence and predicted
limits*/;
output out=resids r=residual;
plot residual.* x;

proc print data=resids;

proc univariate;

var x;
title" Plot of residuals against x";
;
run;
```

Relevant Output:

Parameter Estimates

Variable	DF	Parameter Estimate	Standard Error	t Value	Pr > |t|
Intercept	1	2.42857	0.99933	2.43	0.0511
x	1	0.98810	0.19790	4.99	0.0025

Analysis of Variance

Source	DF	Sum of Squares	Mean Square	F Value	Pr > F
Model	1	41.00595	41.00595	24.93	0.0025
Error	6	9.86905	1.64484		
Corrected Total	7	50.87500			

Root MSE	1.28251	R-Square	0.8060
Dependent Mean	6.87500	Adj R-Sq	0.7737

Descriptive statistics for x

The UNIVARIATE Procedure
Variable: x

Moments

N	8	Sum Weights	8
Mean	4.5	Sum Observations	36
Std Deviation	2.44948974	Variance	6
Skewness	0	Kurtosis	-1.2
Uncorrected SS	204	Corrected SS	42
Coeff Variation	54.4331054	Std Error Mean	0.8660254

Dependent Variable: y

Output Statistics

Obs	Dependent Variable	Predicted Value	Std Error Mean Predict	95% CL Mean		95% CL Predict		Residual
1	2.0000	3.4167	0.8279	1.3910	5.4424	-0.3185	7.1519	-1.4167
2	5.0000	4.4048	0.6711	2.7626	6.0469	0.8629	7.9466	0.5952
3	5.0000	5.3929	0.5420	4.0667	6.7190	1.9860	8.7997	-0.3929
4	8.0000	6.3810	0.4641	5.2453	7.5166	3.0436	9.7183	1.6190
5	9.0000	7.3690	0.4641	6.2334	8.5047	4.0317	10.7064	1.6310
6	7.0000	8.3571	0.5420	7.0310	9.6833	4.9503	11.7640	-1.3571
7	9.0000	9.3452	0.6711	7.7031	10.9874	5.8034	12.8871	-0.3452
8	10.0000	10.3333	0.8279	8.3076	12.3590	6.5981	14.0685	-0.3333

(a) Calculate the least square estimates of the regression line that predicts y from x.

The parameter estimates table of the SAS output of the regression procedure (proc reg) gives the least square estimates of the coefficients of the regression line that predicts y from x. The model is

$$y = \beta_0 + \beta_1 x + \varepsilon$$

where
β_0 = the intercept term, β_1 = the slope of the regression line, and ε = the error term.

From the Parameter Estimates table of the SAS output, it follows that

$$\widehat{\beta_1} = 0.9881$$
$$\widehat{\beta_0} = 2.4286$$

The least square regression line also known as the line of best fit is then:

$$\hat{y} = 2.4286 + 0.9881x$$

(b) Test the hypothesis that $\beta_1 = 0$. Construct a 95% confidence interval on β_1.

The test is performed to determine if there is a linear relationship between x and y. The linear relationship exists if β_1 is significantly different from 0.

Step1: $H_0 : \beta_1 = 0 \qquad H_A : \beta_1 \neq 0$

Step2: $\alpha = 0.05$

Step3: The test statistic is $F^* = \dfrac{MSR}{MSE}$ which under H_0 has an F $(1\text{-}\alpha, 1, n\text{-}2)$ distribution.

Step4: We will reject H_0 if the $F^* > F (0.95, 1, 6) = 5.9874$

Step5: $F^* = \dfrac{41.00595}{1.64484} = 24.93$, $p-value = 0.0025$

Step6: Since $F^* > 5.9874$, we reject H_0.

Note: the F value is just the square of the t value. Therefore, both the F test and T test will lead us to the same conclusion. Also, we could reject H_0 simply by noticing that the p-value = 0.0025 is less than 0.05.

> There is sufficient evidence to support the claim that the slope of the regression line is different from zero. There is a significant linear relationship between x and y.

The 95% confidence interval for β_1 is given by

$$\widehat{\beta_1} \pm t(1 - \alpha/2; n - 2)s(\widehat{\beta_1})$$
$$\equiv 0.9881 \pm (2.447)(0.1979)$$
$$\equiv (0.5038, 1.4724)$$

We are 95% confident that the true value of β_1 is between 0.5038 and 1.4724.

(c) Calculate R^2. Explain it.

The coefficient of determination is $R^2 = 0.8060$. It is obtained from the analysis of variance table of the SAS output of the regression procedure. It is represented as R-Square. We conclude that 80.60% of the variation of y is explained by the regression of y on x. This implies there is a positive, moderate linear relationship between y and x.

(d) Construct a 95% prediction interval on the value of y when $x = 5$.

$$s\{\hat{Y_h}\} = \sqrt{MSE\left[1 + \frac{1}{n} + \frac{(X_h - \overline{X})^2}{\sum_{i=1}^{n}(X_i - \overline{X})^2}\right]}$$

The 95% prediction interval on the value of y when $x = 5$ is then given by:

12

$$\hat{Y}_h \pm t(1-\alpha/2; n-2)s\{\hat{Y}_h\}$$

The confident limits at $x = 5$ can be directly read from the SAS output. The prediction interval is

$$(4.0317, 10.7064)$$

We are 95% confident that the individual value of Y when $X = 5$ will be between 4.0317 and 10.7064.

(e) Calculate the residuals and plot them against x. Explain the plot.

The residuals are as follow:
This table is obtained from the SAS output.

Obs	x	y	residual
1	1	2	-1.41667
2	2	5	0.59524
3	3	5	-0.39286
4	4	8	1.61905
5	5	9	1.63095
6	6	7	-1.35714
7	7	9	-0.34524
8	8	10	-0.33333

The plot of the residuals against x:

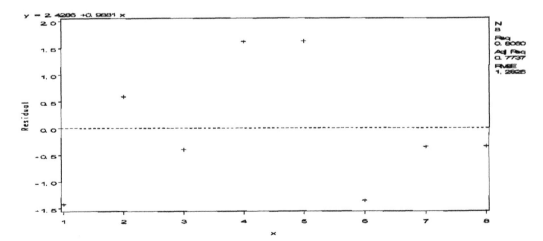

The plot of the residuals against x does not exhibit any discernible pattern. No assumption seems to be violated, that is, there is no evidence of non-constancy of error variance, non-independence of random errors, non-normality of random errors or outliers. Hence, the fitted model is quite good for the given data set. However, the analysis would be more convincing with a larger sample size.

(f) Completely explain the relation between x and y.

$$\hat{y} = 0.9881x + 2.42857$$

The estimated value of y at $x = 0$, also called the y-intercept, is equal to 2.42857. This value is not always interpretable. The slope of the regression line is equal to 0.9881: for every unit increase in x, y will increase by 0.9881. Again, since $R^2 = 0.8060$, 80.60 % of the variation of y is explained by the variation of x.

2. The data in Table 2.9 show the grades for 15 students on the midterm examination and the final average in a statistics course.

SAS Code:

The data for this exercise is reg02p04.

```
proc univariate;
var midterm final;
;
proc reg data = reg02p04;
model final = midterm /noint /*regression through the origin*/;
output out=resids r=residual;
model final = midterm / p cli alpha=.1 clm;
output out=resids2 r=residual;

proc gplot data=resids;
plot residual*midterm;
title "Regression through origin";

proc gplot data=resids2;
plot residual*midterm;
title "Standard Regression";
run;
```

Relevant Output:

Descriptive statistics for x

The UNIVARIATE Procedure
Variable: midterm

Moments

N	15	Sum Weights	15
Mean	73.2	Sum Observations	1098
Std Deviation	17.6764573	Variance	312.457143
Skewness	-0.9437688	Kurtosis	0.07304478
Uncorrected SS	84748	Corrected SS	4374.4
Coeff Variation	24.1481657	Std Error Mean	4.56404165

Fit of regression in part (a)

The REG procedure

Dependent Variable: final

Analysis of Variance

Source	DF	Sum of Squares	Mean Square	F Value	Pr > F
Model	1	2560.59579	2560.59579	19.41	0.0007
Error	13	1715.00421	131.92340		
Corrected Total	14	4275.60000			

Parameter Estimates

Variable	DF	Parameter Estimate	Standard Error	t Value	Pr > |t|
Intercept	1	14.39557	13.05331	1.10	0.2901
midterm	1	0.76509	0.17366	4.41	0.0007

Fit of regression through origin

The REG procedure

Dependent Variable: final

NOTE: No intercept in model. R-Square is redefined.

Analysis of Variance

Source	DF	Sum of Squares	Mean Square	F Value	Pr > F
Model	1	76743	76743	572.87	<.0001
Error	14	1875.45367	133.96098		
Uncorrected Total	15	78618			

Root MSE	11.57415	R-Square	0.9761
Dependent Mean	70.40000	Adj R-Sq	0.9744
Coeff Var	16.44056		

Parameter Estimates

Variable	DF	Parameter Estimate	Standard Error	t Value	Pr > |t|
midterm	1	0.95160	0.03976	23.93	<.0001

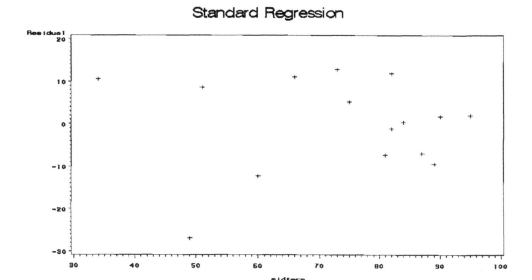

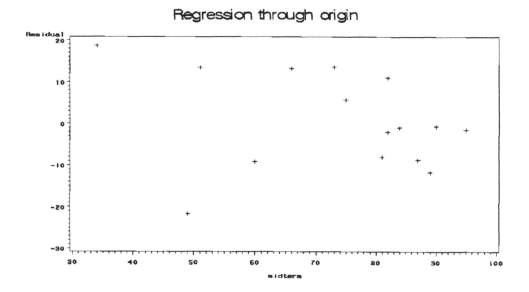

(a) Do the regression analysis to predict the final average based on the midterm examination score.

First, we test the hypothesis that the slope of the regression line is 0

$$final = \beta_0 + \beta_1 midterm + \varepsilon$$

The estimates of the coefficients of the regression line that predicts the final average base on the midterm scores are given in the Parameter Estimates of the SAS output:

$$\widehat{\beta_1} = 0.76509$$

16

$$\widehat{\beta_0} = 14.39557$$

The least-square regression line is then

$$Final = 0.76509 midterm + 14.39557$$

The coefficient of determination R-Square = 0.5989: 59.89% of the variability of the final grade can be explained by the variation of the midterm grade.

Step1: $H_0 : \beta_1 = 0$ vs. $H_A : \beta_1 \neq 0$

Step2: $\alpha = 0.05$

Step3: The test statistic is $F^* = \dfrac{MSR}{MSE}$ has an F $(1-\alpha, 1, 13)$ distribution under H_0.

Step4: We will reject H_0 if the $F^* > F (0.95, 1, 13) = 4.66719$

Step5: $F_1^* = 19.41$, $p-value = 0.0007$

Step6: Since $F_1^* > 4.66719$, we reject H_0.

There is sufficient evidence to conclude that $\beta_1 \neq 0$.

(b) Estimate, using a 90% confidence interval, the value of a midterm score for a student whose final average will be 70.

$$Final = 0.76509 midterm + 14.39557$$

$$\hat{y}_{h(new)} = 70$$

$$\hat{X}_{h(new)} = \frac{\hat{y}_{h(new)} - \hat{\beta}_0}{\hat{\beta}_1} = \frac{70 - 14.39557}{0.76509} = 72.677$$

where X_h denotes variable midterm and Y_h denotes the variable final.

$$s\{X_{pred}\} = \sqrt{\frac{MSE}{\hat{\beta}_1^2}\left[1 + \frac{1}{n} + \frac{(\hat{X}_{h(new)} - \overline{X})^2}{\sum_{i=1}^{n}(X_i - \overline{X})^2}\right]} = \sqrt{\frac{131.9234}{(0.76509)^2}\left[1 + \frac{1}{15} + \frac{(72.677 - 73.2)^2}{4374.4}\right]} = 15.5051$$

Approximate $1-\alpha$ confidence limits for $X_{h(new)}$ are

$$X_{h(new)} \pm t(1-\alpha/2; n-2)s(X_{pred})$$
$$\equiv 72.677 \pm 1.771(15.5051)$$
$$\equiv (45.2, \ 100)$$

We are 90% confident that the midterm score of a student whose final average will be 70 is greater than 45.2.

(c) Fit the regression through the origin and compare it with part (a)

The least-square regression line is

$$Final = 0.9516 midterm$$

The coefficient of midterm is a little larger when there is no intercept. Also, the mean square errors of the two models are almost identical (133.96 and 131.92), implying that the no intercept model provides a better fit of the data than the with-intercept regression. However, it is usually recommended that the intercept term to be used in the analysis since the no-intercept model can lead to wrong results such as the sum of squares due to errors being greater than the total sum of squares. This means that the coefficient of determination, RSQUARE, may turn out to be negative.

5. The *1995 Statistical Abstract of the United States* lists the breakdown of the Consumer Price Index by major groups for the years 1960 to 1994. The Consumer Price Index reflects the buying patterns of all urban consumers. Table 1.20 and file REG02P05 list the major groups energy and transportation.

SAS Code:
The data for this exercise is reg02p05

```
proc corr data = reg02p05;
var Energy Transp Year;

proc reg data = reg02p05;
model Energy = Year;
model Transp = Year;
;
run;
```

Relevant Output:

Pearson Correlation Coefficients, N = 35
Prob > |r| under H0: Rho=0

| | Energy | Transp | Year |

Energy	1.00000	0.97844 <.0001	0.93535 <.0001
Transp	0.97844 <.0001	1.00000	0.96878 <.0001
Year	0.93535 <.0001	0.96878 <.0001	1.00000

Dependent Variable: Energy

Analysis of Variance

Source	DF	Sum of Squares	Mean Square	F Value	Pr > F
Model	1	35672	35672	230.76	<.0001
Error	33	5101.34188	154.58612		
Corrected Total	34	40773			

Root MSE	12.43327	R-Square	0.8749
Dependent Mean	59.86286	Adj R-Sq	0.8711
Coeff Var	20.76958		

Parameter Estimates

Variable	DF	Parameter Estimate	Standard Error	t Value	Pr > |t|
Intercept	1	-183.53695	16.16017	-11.36	<.0001
Year	1	3.16104	0.20809	15.19	<.0001

Dependent Variable: Transp

Analysis of Variance

Source	DF	Sum of Squares	Mean Square	F Value	Pr > F
Model	1	43600	43600	503.90	<.0001
Error	33	2855.34566	86.52563		
Corrected Total	34	46456			

Root MSE	9.30192	R-Square	0.9385
Dependent Mean	70.55714	Adj R-Sq	0.9367
Coeff Var	13.18352		

Parameter Estimates

Variable	DF	Parameter Estimate	Standard Error	t Value	Pr > |t|

```
Intercept     1      -198.53521      12.09018      -16.42      <.0001
Year          1         3.49471       0.15568       22.45      <.0001
```

(a) Perform a correlation analysis to determine the relationship between transportation and energy. Calculate the confidence interval on the correlation coefficient.

The correlation coefficient between transportation and energy is r = 0.97844 which is positive and high. Obviously, the correlation coefficient is significant at any reasonable level of significance. By squaring the correlation coefficient (coefficient of determination), we conclude that about 96% of the variation in energy use can be explained by the variation in transportation. We say that energy and transportation are highly correlated.

To calculate the confidence interval on the correlation coefficient, the following steps are taken.
Using z′ transformation we determine the confidence interval on the correlation coefficient.

$$z' = \frac{1}{2}\ln\left(\frac{1+r_{12}}{1-r_{12}}\right) = \frac{1}{2}\ln\left(\frac{1+0.97844}{1-0.97844}\right) = 2.2596$$

$$\sigma\{z'\} = \frac{1}{\sqrt{n-3}} = \frac{1}{\sqrt{35-3}} = 0.1768$$

Hence, the confidence limits for E [z′] are
$$z' \pm z(1-\alpha/2)\sigma\{z'\}$$
$$\equiv 2.2596 \pm 1.96(0.1768)$$
$$\equiv (1.913, 2.6061)$$
Transforming back to the equation,

$$\rho_{12} = \frac{e^{2E[z']}-1}{e^{2E[z']}+1}$$

we get the confidence interval for correlation coefficient as follows.

$$(0.9573, 0.9892)$$

We are 95 % confident that true value of correlation coefficient between transportation and energy is between 0.9573 and 0.9892. Transportation and energy are then highly correlated.

(b) The estimated regression equation with energy as the dependent variable is:
$\widehat{Energy}$ =-183.54+3.16(Year). This means that for every increase in one year consumption, energy increases on average of 3.16 units. The negative intercept has no

meaning in this problem. The slope for energy is significant. The corresponding t-value is equal to 15.19; the p-value is less than 0.001. Furthermore, the R-square value is 0.8749 indicating that about 87% of the variation in energy consumption can be explained by the regression on the year. The model seems to fit the data well.

The residuals vs. predicted values and residuals vs. Year are plotted.

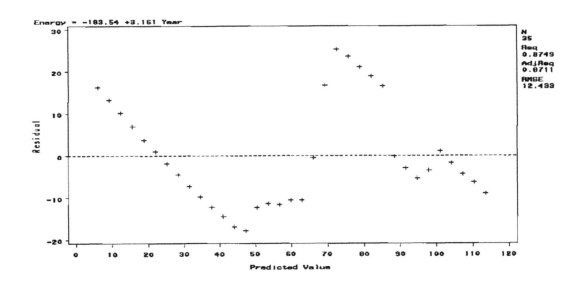

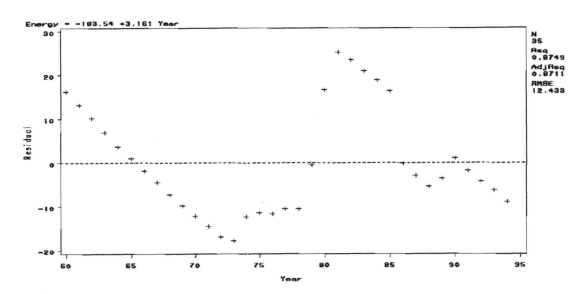

The plots of the residuals vs. predict and the residuals vs. year should be the same (i.e. equivalent), since the predicted value is only a linear transformation of the independent variable for simple linear regression model.

Both the plots seem to indicate a nonlinear relationship between energy consumption and the year. This makes sense because many factors, such as such as weather, and economic conditions contribute to the consumption of energy.

The plot doesn't indicate any outliers, nor does it violate the assumption of homoscedasticity. However, as the data is taken in time sequence, we see that plot of residuals against year shows that at the beginning from 1960 to 1973 and from 1980 to 1995, the values of residuals decrease indicating that there is no strong growth of consumer price index for energy. A linear time related trend effect is observed in the plot of residuals against year. The slope of the line is negative during the years 1960 to 1973 and 1980 to 1995.

The estimated regression equation with transportation as the dependent variable is: $\overline{Trans}$ =-198.54+3.49(Year). This means that for every increase in one year transportation increases on average 3.49 units. The negative intercept has no meaning in this problem. Note that the overall regression is significant. The corresponding F-value is equal to 503.90; the p-value is less than 0.0001. The slope for transportation is also significant. The corresponding t-value is equal to 22.45; the p-value is less than 0.001. (recall that these two tests are identical). Furthermore, the R-square value is 0.9385 indicating that about 94% of the variation in transportation can be explained by the regression on the year. The model seems to fit the data well.

The residuals vs. predicted values and residuals vs. Year are plotted.

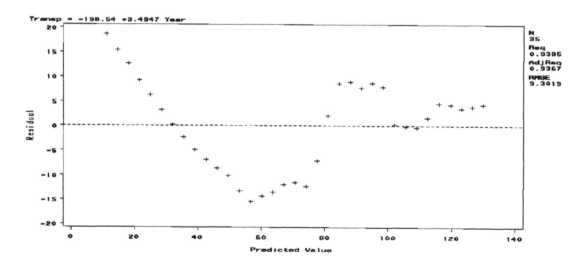

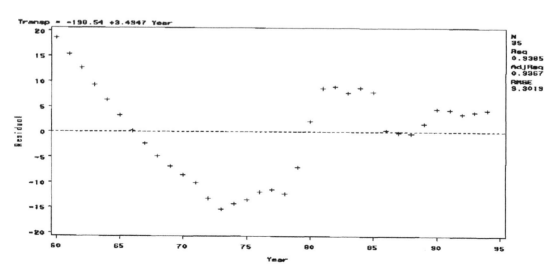

Both the plots seem to indicate a nonlinear relationship between transportation and the year. Again, this makes sense since other factors contribute to transportation such as weather, road condition etc. The plot doesn't indicate the existence of any outliers, nor does it violates the assumption of homoscedasticity. However, if the data is taken in time sequence, the plot does exhibit a time related trend effect. The plot of residuals versus year shows that the magnitudes of residuals is decreasing initially from 1960 -1973 and then increasing from 1973 and 1980, remaining approximately constant to 1995. We can conclude that there is no strong initial growth in consumer price index for transportation between 1960-1973. However, in later periods, the consumer price index appears to increase.

Both plots do indicate that independence in error terms is violated.

Chapter 3
Multiple Linear Regression
Solutions

3. The purpose of this study is to see how individuals' attitudes are influenced by a specific message as well as their previous knowledge of the environment. Respondents were initially tested on their knowledge of the subject (FACT) and were given a test on their attitude (PRE). The sex (SEX) of respondents was also recorded; "1" is female and "2" is male.

The respondents were then exposed to an antipreservation message and the number of positive (NUMPOS) and negative (NUMNEG) reactions recorded. The response variable is POST, a test on preservation attitude.

SAS Code:

The data for this exercise is reg03p03. Again, only the programming code is presented, the input statement and the data are on the cd.

```
proc reg;
model POST = FACT PRE NUMPOS NUMNEG SEX;
proc reg;
model POST = FACT PRE NUMPOS NUMNEG SEX;
output out = out1 p=pred stdp=stdpred;
plot residual.*p.;
proc print data = out1;
var pred stdpred;
run;
```

Relevant Output:

Analysis of Variance

Source	DF	Sum of Squares	Mean Square	F Value	Pr > F
Model	5	246.43847	49.28769	11.95	<.0001
Error	82	338.27743	4.12533		
Corrected Total	87	584.71591			

Root MSE	2.03109	R-Square	0.4215	
Dependent Mean	2.94318	Adj R-Sq	0.3862	
Coeff Var	69.01007			

Parameter Estimates

Variable	DF	Parameter Estimate	Standard Error	t Value	Pr > |t|
Intercept	1	0.80896	1.07092	0.76	0.4522
FACT	1	-0.14536	0.17224	-0.84	0.4012
PRE	1	0.70928	0.09679	7.33	<.0001

NUMPOS	1	0.44662	0.20013	2.23	0.0284
NUMNEG	1	0.13449	0.20411	0.66	0.5118
SEX	1	0.27223	0.47376	0.57	0.5671

Perform a regression to see how the POST score is related to the other variables.

(a) Estimate all relevant parameters and evaluate the fit and appropriateness of the model.

The least square estimates of the regression line that predicts POST from the other variables, is obtained from the SAS output above.
The least square regression line is:

$$\widehat{POST} = 0.809 - 0.145 FACT + 0.709 PRE + 0.447 NUMPOS + 0.134 NUMNEG + 0.272 SEX$$

The overall model is significant. The corresponding F statistics has a p-value less than 0.0001. The R-sq is only 0.42, indicating that other variables may be influencing the response.

(b) Interpret the coefficients, and their statistical and practical significance.

A one-unit increase on the knowledge of the subject, FACT, is associated with a decrease of 0.145 on preservation attitude, POST, holding constant PRE, NUMPOS, NUMNEG, and SEX.
A one-unit increase on the test of attitude, PRE, is associated with an increase of 0.709 on preservation attitude, POST, holding constant FACT, NUMPOS, NUMNEG, and SEX.
A one-unit increase on the number of positive reactions, NUMPOS, is associated with an increase of 0.447 on preservation attitude, POST, holding constant FACT, PRE, NUMNEG, and SEX.
A one-unit increase on the number of negative reactions, NUMNEG, is associated with an increase of 0.134 on preservation attitude, POST, holding constant FACT, PRE, NUMPOS, and SEX.
Male individuals tend to have a have a higher preservation attitude, POST, than female individuals by an average of 0.272, holding constant FACT, PRE, NUMPOS, and NUMNEG.

The p-values associated to the intercept and most of the predictor variables are large.
The only significant coefficients are the test of attitude (PRE) associated with a p-value of less than 0.0001 and the number of positive reactions (NUMPOS) associated with a p-value of 0.0284. However, there is no indication that the model should be changed at this time.

(c) Examine residuals for possible violations of assumptions.

The plot of residuals vs. predicted below indicates that there may be a problem of non-constant variance. Further investigation might be warranted.

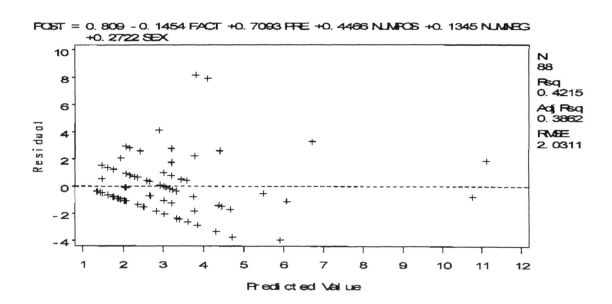

POST = 0.809 - 0.1454 FACT +0.7093 PRE +0.4466 NUMPOS +0.1345 NUMNEG +0.2722 SEX

N 88
Rsq 0.4215
Adj Rsq 0.3862
RMSE 2.0311

(d) The predicted values and their standard errors are listed below:

Obs	pred	stdpred
1	1.4813	0.44070
2	6.7261	0.70610
3	2.3745	0.34878
4	1.6266	0.43729
5	3.0234	0.41303
6	3.2291	0.35340
7	3.1688	0.40447
8	2.1797	0.39880
9	4.7327	0.54952
10	1.7611	0.39543
11	6.0847	0.65310
12	1.4780	0.41938
13	2.2924	0.70681
14	2.9279	0.42103
15	2.9061	0.38385
16	3.1656	0.56926
17	2.4486	0.49201
18	1.4889	0.46393
19	2.0700	0.42481
20	4.6767	0.51358
21	11.1172	1.26700
22	3.0697	0.54124
23	4.1128	0.49884
24	1.4780	0.41938
25	3.0419	0.41258
26	3.2183	0.35637
27	3.7717	0.47180
28	5.9209	0.51468

29	1.6266	0.43729
30	5.5023	0.54028
31	2.5304	0.59090
32	1.8739	0.63922
33	2.5090	0.36376
34	3.4713	0.43104
35	2.0914	0.47098
36	3.6058	0.39345
37	1.8924	0.64719
38	3.8597	0.80262
39	1.4813	0.44070
40	1.9464	0.51083
41	3.0343	0.36631
42	1.4813	0.44070
43	2.6652	0.44933
44	3.3541	0.35659
45	1.7825	0.41652
46	3.7931	0.35034
47	3.6167	0.43414
48	2.9460	0.38489
49	4.3260	0.46587
50	4.4605	0.46433
51	1.7825	0.41652
52	10.7695	1.16335
53	3.7835	0.45595
54	1.6234	0.44697
55	4.4035	0.64594
56	2.0732	0.34873
57	2.0732	0.34873
58	3.2196	0.42793
59	1.6339	0.51630
60	2.0838	0.52797
61	1.6234	0.44697
62	2.6758	0.49767
63	3.2582	0.51210
64	3.4131	0.62721
65	1.3544	0.56035
66	1.4813	0.44070
67	2.6125	0.59827
68	2.1719	0.93192
69	1.8998	0.60214
70	1.7621	0.60596
71	1.6371	0.53706
72	3.8211	0.54352
73	3.2291	0.35340
74	2.0515	0.39301
75	2.5199	0.36346
76	3.0234	0.41303
77	1.6266	0.43729
78	2.8396	0.57697
79	2.0406	0.55381
80	1.4454	0.77529
81	2.0624	0.30457
82	1.3649	0.68229
83	1.4813	0.44070
84	2.7048	0.78286
85	2.3745	0.34878

86	4.3850	0.48023
87	1.9355	0.40014
88	3.3359	0.60561

(e) Summarize the results, including possible recommendations for additional analyses. Obviously the number of independent variables should be systematically evaluated, and other possible variables identified and measured. The lack of constant variance should also be examined and perhaps a transformation is warranted.

5. The data in Table 3.11 are Consumer price Index values. The two groups, energy (ENERGY) and transportation (TRANS) were used in Exercise5 of Chapter 2. Added to these two groups are the index for medical care (MED) and the average for all items (ALL).

SAS Code:

The data for this exercise is reg03p05.
```
proc reg ;
model ALL =  ENERGY TRANS MED ;
plot residual.*p.;
run;
```

Relevant Output:

Dependent Variable: ALL

Analysis of Variance

Source	DF	Sum of Squares	Mean Square	F Value	Pr > F
Model	3	56280	18760	15094.9	<.0001
Error	31	38.52663	1.24279		
Corrected Total	34	56318			

Root MSE	1.11481	R-Square	0.9993
Dependent Mean	74.00857	Adj R-Sq	0.9992
Coeff Var	1.50632		

Parameter Estimates1

Variable	DF	Parameter Estimate	Standard Error	t Value	Pr > |t|
Intercept	1	-3.84289	1.53493	-2.50	0.0178
ENERGY	1	-0.34794	0.08043	-4.33	0.0001
TRANS	1	1.34880	0.13158	10.25	<.0001
MED	1	0.04412	0.03893	1.13	0.2658

ALL vs ENERGY:

Analysis of Variance

Source	DF	Sum of Squares	Mean Square	F Value	Pr > F
Model	1	51848	51848	382.77	<.0001
Error	33	4470.05912	135.45634		
Corrected Total	34	56318			

Root MSE	11.63857	R-Square	0.9206	
Dependent Mean	74.00857	Adj R-Sq	0.9182	
Coeff Var	15.72598			

Parameter Estimates

Variable	DF	Parameter Estimate	Standard Error	t Value	Pr > |t|
Intercept	1	6.50361	3.97183	1.64	0.1110
ENERGY	1	1.12766	0.05764	19.56	<.0001

ALL vs. TRANS:

Analysis of Variance

Source	DF	Sum of Squares	Mean Square	F Value	Pr > F
Model	1	55951	55951	5023.29	<.0001
Error	33	367.56133	11.13822		
Corrected Total	34	56318			

Root MSE	3.33740	R-Square	0.9935	
Dependent Mean	74.00857	Adj R-Sq	0.9933	
Coeff Var	4.50947			

Parameter Estimates

Variable	DF	Parameter Estimate	Standard Error	t Value	Pr > |t|
Intercept	1	-3.42405	1.22957	-2.78	0.0088
TRANS	1	1.09745	0.01548	70.88	<.0001

ALL vs MED:

Analysis of Variance

Source	DF	Sum of Squares	Mean Square	F Value	Pr > F
Model	1	54424	54424	948.23	<.0001
Error	33	1894.04776	57.39539		
Corrected Total	34	56318			

Root MSE	7.57597	R-Square	0.9664	
Dependent Mean	74.00857	Adj R-Sq	0.9653	
Coeff Var	10.23662			

Parameter Estimates

Variable	DF	Parameter Estimate	Standard Error	t Value	Pr > |t|
Intercept	1	19.57559	2.18279	8.97	<.0001
MED	1	0.68364	0.02220	30.79	<.0001

(a) Do the multiple regression using ALL as the dependent variable and the three others as independent variables. The interpretation of the coefficients is quite important here. What do the residuals imply here?

The least square estimates of the regression line that predicts ALL from the other variables is:

$$\widehat{ALL} = -3.843 - 0.348\,ENERGY + 1.349\,TRANS + 0.044\,MED$$

The overall model is significant. The p-value associated with the model is less than 0.0001. Also, the coefficient of determination R-square is equal to 0.9993 meaning that 99.93% of the variability of ALL can be explained by the regression on the three independent variables ENERGY, TRANS, and MED. The intercept term of the regression line is not interpretable since the data set does not contain values of zero for all the independent variables. It further does not make sense as all must be positive.

A one-unit increase of ENERGY is associated with a decrease of 0.348 on the average of all items, ALL, holding the remaining predictor variables constant.

A one-unit increase of TRANS is associated with an increase of 1.349 on the average of all items, ALL, holding the remaining predictor variables constant.

A one-unit increase of MED, is associated with an increase of 0.044 on the average of all items, ALL, holding the remaining predictor variables constant.

The residuals do exhibit any clear pattern. However, there does seem to be a cyclic pattern that could be the time effect not considered in the model.

30

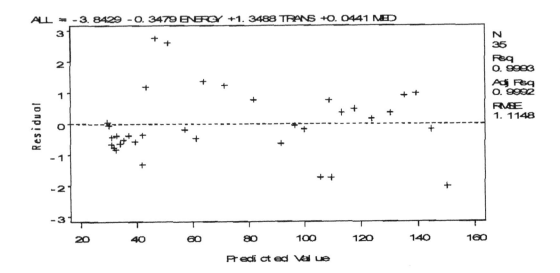

ALL = -3.8429 - 0.3479 ENERGY +1.3488 TRANS +0.0441 MED

N
35

Rsq
0.9993

Adj Rsq
0.9992

RMSE
1.1148

(b) Do separate simple linear regression using ALL as the dependent variable and the other three separately as independent variables. Compare with (a). Explain

The individual simple linear regressions of the response variable ALL against each of the independent variables ENERGY, TRANS, and MED yield the following:

$$ALL = 6.5036 + 1.1277 ENERGY$$
$$ALL = -3.4241 + 1.0974 TRANS$$
$$ALL = 19.576 + 0.6836 MED$$

Note that the partial coefficients for all the variables differ markedly from the total coefficients. In fact the coefficient for ENERGY has a different sign. This is probably due to the fact that all the independent variables are correlated, a condition that will be examined in Chapter 5.

(c) Find the simultaneous confidence region on all three coefficients in part (a) using Bonferroni method.

The $1 - \alpha$ simultaneous confidence region on all three coefficients is given by the Bonferroni procedure:

$$\widehat{\beta_i} \pm Bs\{\widehat{\beta_i}\}; \ B = t(1 - \alpha / 2r)$$

where r is the number of intervals to be computed. If we wish to make simultaneous 90% intervals for β_1, β_2 and β_3, we will use $B = t(1 - \alpha / 2r) = t_{0.0167}(31) \approx 2.226$. The three confidence intervals that make up the family of confidence intervals are:

$-0.3479 \pm (2.226)(0.0804)$, or $(-0.5269, -0.1689)$

$1.3488 \pm (2.226)(0.1316)$, or $(1.0559, 1.6417)$

$0.0441 \pm (2.226)(0.0389)$, or $(-0.0425, 0.1307)$

We conclude that β_1 is between -0.5269 and -0.1689, *and* β_2 is between 1.0559 and 1.6417, *and* β_3 is between -0.0425 and 0.1307, with a family confidence interval of 90%.

(d) Find the individual confidence intervals in part (b) and compare with the results in part (c)

The $1 - \alpha$ individual confidence region for all three coefficients is given by:

$\widehat{\beta_i} \pm t(1 - \alpha/2)s\{\widehat{\beta_i}\}$ where $t(1 - \alpha/2) = t_{0.05}(33) \approx 1.692$

The three confidence intervals are:

$1.1277 \pm (1.692)(0.0576) \equiv (1.0302, 1.2252)$. We are 90% confident that β_1 is between 1.0302 and 1.2252.

$1.0975 \pm (1.692)(0.0155) \equiv (1.0713, 1.1237)$. We are 90% confident that β_2 is between 1.0713 and 1.1237.

$0.6836 \pm (1.692)(0.0222) \equiv (0.6460, 0.7212)$. We are 90% confident that β_3 is between 0.6460 and 0.7212.

9. The data for this study relate certain factors to gasoline consumption. The variables are:

STATE: Two-character aabbreviation
GAS: Total gasoline and auto diesel consumption
AREA: Of state in 1000 miles
POP: Population in millions
MV: Estimated number of registered vehicles in millions
INC: Personal income in billions of dollars
VAL: Value added by manufacturers in billions of dollars
Region: Codes for East or West of the Mississippi river

SAS Code:

The data for this exercise is reg03p09.

```
proc glm;
model gas = area pop mv inc val;
run;
```

Relevant Output:

Analysis of Variance

Source	DF	Sum of Squares	Mean Square	F Value	Pr > F
Model	5	3176631	635326	1371.26	<.0001
Error	42	19459	463.31503		
Corrected Total	47	3196091			

Variable	DF	Parameter Estimate	Standard Error	t Value	Pr > |t|
Root MSE		21.52475	R-Square	0.9939	
Dependent Mean		277.23896	Adj R-Sq	0.9932	
Coeff Var		7.76397			

Parameter Estimates

Variable	DF	Parameter Estimate	Standard Error	t Value	Pr > |t|
Intercept	1	-14.27654	6.08037	-2.35	0.0237
AREA	1	0.21732	0.08231	2.64	0.0116
POP	1	15.96010	8.67583	1.84	0.0729
MV	1	105.85010	7.55905	14.00	<.0001
INC	1	-2.73949	1.77384	-1.54	0.1300
VAL	1	-1.70694	1.68123	-1.02	0.3158

The least square regression line considering all the available predictor variables is:

$$\widehat{gas} = -14.277 + 0.217\,area + 15.960\,pop + 105.850\,mv - 2.739\,inc - 1.707\,val$$

The overall model is significant; the p-value is less than 0.0001. The coefficients corresponding to the variables area and mv are significant with p-values of 0.0013 and less than 0.0001 respectively. The R-Square is 0.99 which indicates that the model basically explains all of the variation in gas consumption.

The residuals plotted below apparently do not have a discernable pattern indicating that there are no obvious violations of assumptions.

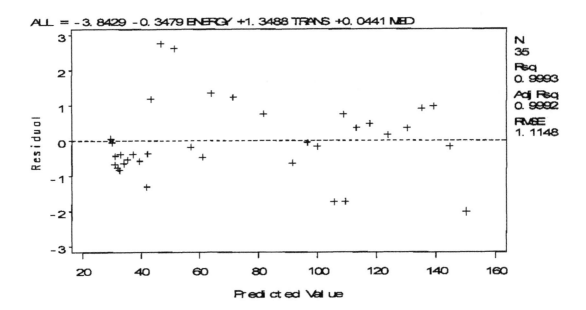

The regression equation fits the data quite good, and the high R-Square indicates that there are no other variables that have a large effect on gas consumption. Additional analysis might reveal that the different regions have different relationships, especially the regions with a lot of mountains. It should be possible to reduce the number of independent variables, especially since only two of the partial regression coefficients are significant. This procedure could be done using the tools described in Chapter 6.

Chapter 4
Problems with Observations
Solutions

3. The Galapagos Islands are a source of data for various types of biological studies. One such study relates the number of plant species to various characteristics of the island. The variables are:

ISLAND: The "official name" of the island in English. Note that many maps give other names to the islands.

AREA: Area in square miles

HEIGHT: Maximum elevation in feet above mean sea level

DSNEAR: Distance to the nearest island in miles

DCENT: Distance to the center of the archipelago

ARNEAR: Area of nearest island in square miles

SPECIES: Number of plant species

SAS Code:

The data for this exercise is reg04p03.

```
proc reg;
id island;
model species = AREA HEIGHT DSNEAR DCENT ARNEAR/influence r ;
plot residual.*p.;
run;
```

Relevant Output:

Dependent Variable: SPECIES

Analysis of Variance

Source	DF	Sum of Squares	Mean Square	F Value	Pr > F
Model	5	164663	32933	11.58	0.0004
Error	11	31280	2843.64619		
Corrected Total	16	195944			

Root MSE	53.32585	R-Square	0.8404	
Dependent Mean	119.29412	Adj R-Sq	0.7678	
Coeff Var	44.70116			

Parameter Estimates

Variable	DF	Parameter Estimate	Standard Error	t Value	Pr > |t|

Intercept	1	16.48765	31.65058	0.52	0.6127
AREA	1	-0.02995	0.04335	-0.69	0.5039
HEIGHT	1	0.07250	0.01678	4.32	0.0012
DSNEAR	1	3.98019	1.36261	2.92	0.0139
DCENT	1	-1.07454	0.33153	-3.24	0.0079
ARNEAR	1	-0.09685	0.03443	-2.81	0.0169

Obs	ISLAND	Dependent Variable	Predicted Value	Std Error Mean Predict	Residual	Std Error Residual	Student Residual
1	Culpepper	7.0000	-24.2956	37.1912	31.2956	38.216	0.819
2	Wenman	14.0000	13.5285	30.4456	0.4715	43.780	0.0108
3	Tower	22.0000	88.6826	29.5868	-66.6826	44.365	-1.503
4	Jervis	42.0000	48.8250	22.4662	-6.8250	48.362	-0.141
5	Bindloe	47.0000	93.6535	14.8623	-46.6535	51.213	-0.911
6	Barrington	48.0000	76.4005	22.2330	-28.4005	48.470	-0.586
7	Gardiner	48.0000	-18.6320	28.1107	66.6320	45.315	1.470
8	Seymour	52.0000	15.9457	29.5825	36.0543	44.368	0.813
9	Hood	79.0000	123.7564	25.4360	-44.7564	46.868	-0.955
10	Narborough	80.0000	95.2486	52.0643	-15.2486	11.531	-1.322
11	Duncan	103.0000	106.5154	19.7633	-3.5154	49.528	-0.0710
12	Abingdon	119.0000	168.3030	29.1391	-49.3030	44.660	-1.104
13	Indefatigable	193.0000	211.1834	28.5205	-18.1834	45.058	-0.404
14	James	224.0000	225.0827	31.2896	-1.0827	43.181	-0.0251
15	Chatham	306.0000	259.1407	29.3193	46.8593	44.542	1.052
16	Charles	319.0000	219.6116	30.7900	99.3884	43.539	2.283
17	Albemarle	325.0000	325.0500	52.8550	-0.0500	7.070	-0.0071

Output Statistics

Obs	ISLAND	-2-1 0 1 2	Cook's D	RStudent	Hat Diag H	Cov Ratio	DFFITS
1	Culpepper	| |* |	0.106	0.8058	0.4864	2.3650	0.7841
2	Wenman	| | |	0.000	0.0103	0.3260	2.6281	0.0071
3	Tower	| ***| |	0.167	-1.6077	0.3078	0.6443	-1.0721
4	Jervis	| | |	0.001	-0.1347	0.1775	2.1306	-0.0626
5	Bindloe	| *| |	0.012	-0.9033	0.0777	1.1997	-0.2621
6	Barrington	| *| |	0.012	-0.5676	0.1738	1.7728	-0.2604
7	Gardiner	| |** |	0.139	1.5641	0.2779	0.6599	0.9703
8	Seymour	| |* |	0.049	0.7992	0.3077	1.7651	0.5328
9	Hood	| *| |	0.045	-0.9508	0.2275	1.3645	-0.5160
10	Narborough	| **| |	5.942	-1.3749	0.9532	13.4075	-6.2080
11	Duncan	| | |	0.000	-0.0677	0.1374	2.0480	-0.0270
12	Abingdon	| **| |	0.086	-1.1162	0.2986	1.2486	-0.7283
13	Indefatigable	| | |	0.011	-0.3877	0.2860	2.2689	-0.2454
14	James	| | |	0.000	-0.0239	0.3443	2.7008	-0.0173
15	Chatham	| |** |	0.080	1.0577	0.3023	1.3439	0.6962
16	Charles	| |**** |	0.434	3.0003	0.3334	0.0565	2.1217
17	Albemarle	| | |	0.000	-0.006737	0.9824	100.7692	-0.0504

Obs	ISLAND	-2-1 0 1 2	D	RStudent	H	Ratio	DFFITS

1	Culpepper	| |* |	0.106	0.8058	0.4864	2.3650	0.7841
2	Wenman	| | |	0.000	0.0103	0.3260	2.6281	0.0071
3	Tower	| ***| |	0.167	-1.6077	0.3078	0.6443	-1.0721
4	Jervis	| | |	0.001	-0.1347	0.1775	2.1306	-0.0626
5	Bindloe	| *| |	0.012	-0.9033	0.0777	1.1997	-0.2621
6	Barrington	| *| |	0.012	-0.5676	0.1738	1.7728	-0.2604
7	Gardiner	| |** |	0.139	1.5641	0.2779	0.6599	0.9703
8	Seymour	| |* |	0.049	0.7992	0.3077	1.7651	0.5328
9	Hood	| *| |	0.045	-0.9508	0.2275	1.3645	-0.5160
10	Narborough	| **| |	5.942	-1.3749	0.9532	13.4075	-6.2080
11	Duncan	| | |	0.000	-0.0677	0.1374	2.0480	-0.0270
12	Abingdon	| **| |	0.086	-1.1162	0.2986	1.2486	-0.7283
13	Indefatigable	| | |	0.011	-0.3877	0.2860	2.2689	-0.2454
14	James	| | |	0.000	-0.0239	0.3443	2.7008	-0.0173
15	Chatham	| |** |	0.080	1.0577	0.3023	1.3439	0.6962
16	Charles	| |**** |	0.434	3.0003	0.3334	0.0565	2.1217
17	Albemarle	| | |	0.000	-0.006737	0.9824	100.7692	-0.0504

		--------------------------------DF8ETAS--------------------------					
Obs	ISLAND	Intercept	AREA	HEIGHT	DSNEAR	DCENT	ARNEAR
1	Culpepper	-0.0713	0.0B04	-0.0814	-0.1544	0.6865	-0.0287
2	Wenman	-0.0006	0.0003	-0.0004	-0.0010	0.0059	-0.0006
3	Tower	-0.2116	-0.4617	0.5764	-0.7548	0.2932	-0.4009
4	Jervis	-0.0606	-0.0089	0.0295	0.0241	0.0187	-0.0102
5	Bindloe	-0.1381	0.0340	0.0063	0.0387	0.0290	0.0755
6	Barrington	-0.2026	-0.0657	0.1433	-0.0267	0.1451	-0.1195
7	Gardiner	0.7873	0.1483	-0.4024	-0.6934	0.2928	-0.0198
8	Seymour	0.5041	0.1519	-0.3286	-0.1862	-0.1791	0.2049
9	Hood	-0.0521	-0.1291	0.1622	-0.3904	0.1573	-0.0960
10	Narborough	1.3471	0.6076	-0.7540	0.2325	-1.0259	-3.7604
11	Duncan	-0.0200	0.0048	0.0026	0.0052	0.0140	-0.0030
12	Abingdon	0.1816	0.5270	-0.5902	0.1636	-0.2649	0.4971
13	Indefatigable	-0.0751	0.1150	-0.1399	0.1099	0.0472	0.1614
14	James	-0.0015	0.0119	-0.0132	0.0057	0.0022	0.0127
15	Chatham	-0.2530	-0.25B9	0.3870	0.4454	-0.2045	-0.2699
16	Charles	-0.5066	-0.2551	0.3554	1.8080	-1.0596	0.2853
17	Albemarle	-0.0003	-0.0335	0.0069	-0.0009	-0.0015	-0.0028

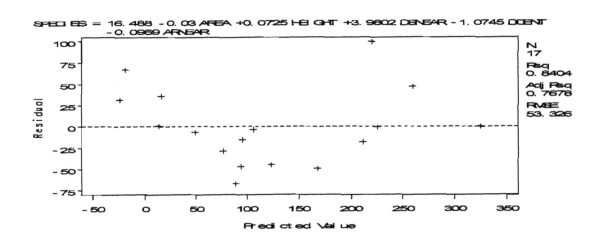

SPECIES = 16.488 - 0.03 AREA + 0.0725 HEIGHT + 3.9802 DSNEAR - 1.0745 DCENT - 0.0969 ARNEAR

N 17
Rsq 0.8404
Adj Rsq 0.7678
RMSE 53.326

Perform a regression to estimate the number of species. Look for outliers and influential observation(s).

From the SAS output the regression is:

$$species = 16.488 - 0.030\,area + 0.073\,height + 3.980\,dsnear - 1.075\,dcent - 0.097\,arnear$$

None of the studentized residuals has a value greater than 2.5. and the residual plot does not indicate any severe potential outliers. Therefore, there does not appear to be any outliers. For the detection of leverage points, we need to find the diagonals of the hat matrix that have values higher than $2\dfrac{m+1}{n} = 2\dfrac{5+1}{17} \approx 0.706$. Thus, possible leverage points are observations 10 and 17.

The approximate standard error of DFFITS is $\sqrt{\dfrac{m+1}{n}}$ where m is the number of predictor variables and n is the number of observations. In this case, $\sqrt{\dfrac{m+1}{n}} = \sqrt{\dfrac{5+1}{17}} = 0.594$.
Thus, we need to find DFFITS (absolute value) of observations exceeding 2(0.594) =1.188. This occurs with observation 10 and observation 16 (Also, the values of DFBETAS (absolute value) for observations 10 and 16 are greater than $\dfrac{2}{\sqrt{n}} = \dfrac{2}{\sqrt{17}} \approx 0.485$. These values for DFBETAS are then considered large.
Since only one observation satisfies all conditions, we conclude that the only leverage point is observation 10.

To determine if it is an influential observation we look at the COVRATIO statistic. The COVRATIO statistics of observation 10 is equal to 13.4075. Since this value does not belong to the interval $1 \pm 3(m+1)/n \equiv 1 \pm 3(5+1)/17 \cong (-0.059, 2.059)$, we conclude that observation 10 is an influential observation.
Note that observation 10 is the island Narborough which is one of the largest islands close to other islands, but not close to the center of the archipelago. It appears to have fewer species than expected.

5. The data for this exercise contains sorghum grain prices received by farmers in a South Texas county from 1980 through 1987. The variables are:
   N: Sequential month number from 1 thorugh 96
   YEAR: The actual year
   MONTH: Labeled 1 through 12
   PRICE: Dollars per bushel

SAS Code
The data for this exercise is reg04p05.

38

```
proc autoreg ;
model PRICE = N / nlag = 1 dwprob;
run;
```

Relevant Output:

The AUTOREG Procedure

Dependent Variable   PRICE

Ordinary Least Squares Estimates

SSE	12.7484827	DFE	94
MSE	0.13562	Root MSE	0.36827
SBC	87.7470454	AIC	82.618349
Regress R-Square	0.6179	Total R-Square	0.6179
Durbin-Watson	0.1633	Pr < DW	<.0001
Pr > DW	1.0000		

NOTE: Pr<DW is the p-value for testing positive autocorrelation, and Pr>DW is the p-value for testing negative autocorrelation.

Variable	DF	Estimate	Standard Error	t Value	Approx Pr > |t|
Intercept	1	3.3646	0.0758	44.41	<.0001
N	1	-0.0167	0.001356	-12.33	<.0001

Estimates of Autocorrelations

Lag	Covariance	Correlation	-1 9 8 7 6 5 4 3 2 1 0 1 2 3 4 5 6 7 8 9 1
0	0.1328	1.000000	| |********************|
1	0.1200	0.903328	| |****************** |

Preliminary MSE   0.0244

Estimates of Autoregressive Parameters

Lag	Coefficient	Standard Error	t Value
1	-0.903328	0.044480	-20.31

Yule-Walker Estimates

SSE	2.02250118	DFE	93
MSE	0.02175	Root MSE	0.14747
SBC	-82.739199	AIC	-90.432243
Regress R-Square	0.1046	Total R-Square	0.9394
Durbin-Watson	1.3468	Pr < DW	0.0003
Pr > DW	0.9997		

NOTE: Pr<DW is the p-value for testing positive autocorrelation, and Pr>DW is the p-value for testing negative autocorrelation.

The AUTOREG Procedure

	Standard	Approx

Variable	DF	Estimate	Error	t Value	Pr > |t|
Intercept	1	3.1898	0.2531	12.60	<.0001
N	1	-0.0142	0.004313	-3.30	0.001

Fit a linear trend using Price as the dependent variable and N as the independent variable. Check for autocorrelation. Redo the analysis if necessary.
The least square regression line is

$$\text{Pr} ice = 3.365 - 0.017N .$$

The autocorrelation is certainly significant. Ordinary least squares estimates show a highly significant coefficient of -0.016 with a coefficient of determination of 0.62. The Yule Walker estimates show a coefficient of -0.014 with, however, a much larger p-value. Furthermore, the coefficient of determination due to the structural model is shown to be 0.10 while the one due to autocorrelation is the difference between the total coefficient of determination and the coefficient of determination due to the structural model: $0.94 - 0.10 = 0.84$
The new estimated model is

$$\text{Pr} ice = 3.190 - 0.014N .$$

Note: The proc autoreg with nlag =3 shows that the lag 2 and lag 3 are not significant. Therefore, nlag=1 is sufficient.

Chapter 5
Multicollinearity
Solutions

1. The data is from a set of tournaments featuring professional putters. The variables are:
   TNMT: The number of tournaments in which the player participated
   WINS: The number of tournaments won
   AVGMON: Average money won per tournament
   PNTS: The number of "Player of the Year" points.
   ASA: The player's adjusted point average
   PRICE:the "price" placed on the player at the end of the tour.

SAS code:
The data for this exercise is reg05p01. Only the SAS programming commands are given. The input statements and the data are given in the data sets on the cd accompanying the text.

```
proc reg data = reg05p01;
model price = TNMT     WINS     AVGMON     PNTS     ASA  /vif collin;
;
Proc Princomp data = reg05p01 out = prin;
var TNMT     WINS     AVGMON     PNTS     ASA     ;
;
proc reg data = prin ;
model price = prin1- prin5;
;
run;
```

Relevant Output:

Analysis of Variance

Source	DF	Sum of Squares	Mean Square	F Value	Pr > F
Model	5	45694	9138.81380	40.55	<.0001
Error	26	5859.93100	225.38196		
Corrected Total	31	51554			

Root MSE	15.01273	R-Square	0.8863	
Dependent Mean	53.50000	Adj R-Sq	0.8645	

Parameter Estimates

Variable	DF	Parameter Estimate	Standard Error	t Value	Pr > |t|	Variance Inflation
Intercept	1	663.62369	270.29149	2.46	0.0211	0
TNMT	1	0.65681	0.74771	0.88	0.3878	2.85816
WINS	1	-0.04444	6.07845	-0.01	0.9942	6.14237

AVGMON	1	-0.09666	0.24485	-0.39	0.6962	7.50031
PNTS	1	0.82093	0.35596	2.31	0.0293	15.76682
ASA	1	-20.64129	8.51398	-2.42	0.0226	6.36166

Collinearity Diagnostics (intercept adjusted)

Number	Eigenvalue	Condition Index
1	3.61834	1.00000
2	0.98714	1.91454
3	0.25626	3.75760
4	0.09450	6.18799
5	0.04375	9.09384

Collinearity Diagnostics (intercept adjusted)

-----------------------Proportion of Variation----------------------

Number	TNMT	WINS	AVGMON	PNTS	ASA
1	0.00448	0.01046	0.00892	0.00456	0.00951
2	0.29397	0.00029696	0.00781	0.00159	0.01373
3	0.04927	0.35972	0.00310	0.00007126	0.24012
4	0.04577	0.09696	0.97875	0.03653	0.30484
5	0.60652	0.53256	0.00142	0.95725	0.43180

The PRINCOMP Procedure

Simple Statistics

	TNMT	WINS	AVGMON	PNTS	ASA
Mean	13.84375000	0.781250000	44.09375000	33.37500000	31.11812500
StD	6.09658485	1.099394995	30.15910431	30.07812408	0.79878714

Correlation Matrix

	TNMT	WINS	AVGMON	PNTS	ASA
TNMT	1.0000	0.3028	0.1796	0.5254	-.1304
WINS	0.3028	1.0000	0.8354	0.8786	-.7234
AVGMON	0.1796	0.8354	1.0000	0.8512	-.8816
PNTS	0.5254	0.8786	0.8512	1.0000	-.8123
ASA	-.1304	-.7234	-.8816	-.8123	1.0000

Eigenvalues of the Correlation Matrix

	Eigenvalue	Difference	Proportion	Cumulative
1	3.61834253	2.63119851	0.7237	0.7237
2	0.98714402	0.73087972	0.1974	0.9211
3	0.25626431	0.16176893	0.0513	0.9724
4	0.09449538	0.05074161	0.0189	0.9912
5	0.04375377		0.0088	1.0000

Eigenvectors

	Prin1	Prin2	Prin3	Prin4	Prin5
TNMT	0.215169	0.910718	-.189959	0.111183	-.275406
WINS	0.482262	-.042434	0.752475	-.237227	-.378322
AVGMON	0.491978	-.240530	-.077142	0.832878	-.021569
PNTS	0.510095	0.157246	0.016968	-.233283	0.812630
ASA	-.467853	0.293616	0.625666	0.428082	0.346686

The REG Procedure
Dependent Variable: PRICE
Analysis of Variance

Source	DF	Sum of Squares	Mean Square	F Value	Pr > F
Model	5	45694	9138.81380	40.55	<.0001
Error	26	5859.93100	225.38196		
Corrected Total	31	51554			

Root MSE	15.01273	R-Square	0.8863	
Dependent Mean	53.50000	Adj R-Sq	0.8645	
Coeff Var	28.06117			

Parameter Estimates

Variable	DF	Parameter Estimate	Standard Error	t Value	Pr > |t|	
Intercept	1	53.50000	2.65390	20.16	<.0001	
Prin1	1	19.71314	1.41750	13.91	<.0001	
Prin2	1	3.39161	2.71387	1.25	0.2225	835
Prin3	1	-10.46954	5.32641	-1.97	0.0601	
Prin4	1	-14.78956	8.77149	-1.69	0.1037	
Prin5	1	13.32794	12.89054	1.03	0.3107	

The first output from Proc Reg, indicates that the overall regression is significant and the R-Square =.88, implying a good fit, even though PTS and ASA have the only significant partial regression coefficients. The VIF values seem to indicate the existence of multicollinearity, especially concentrated on the Points variable.

1. Determine the extent and nature of multicollinearity.

The variance inflation factor (VIF) for the model is VIF_{model}

$$= \frac{1}{1-R^2} \approx \frac{1}{1-0.8863} \approx 8.795.$$

The VIF for PNTS is higher than the VIF for the model. The VIF for AVGMON and ASA are smaller but close to the VIF_{model} which indicate possible multicollinearity problems. The correlation matrix indicates relative high correlations among the three predictors PNTS, ASA, and AVGMON. Examining the eigenvalues, and the variance proportions, it appears that two components explain over 92% of the variation. The first component is a function of all variables (with ASA being a negative contributor). The second is primarily a function of tournaments played.

2. Discuss possible practical reasons why the multicollinearity exists

The multicollinearity reflects the fact that the variance of the variable Price of the player should be based on all the measurements of quality of play. The ASA variable contributes negatively reflecting the effect of harder courses on the other variables. While the second component points at a reasonable assumption (the more tournaments you enter the more you are likely to win) we will see that this relationship is not really important.

3. Implement at least one form of remedial action and determine if these methods provide more useful results

The last output from Proc Reg uses the principal components as independent variables. All the model statistics are exactly the same as the original regression, however only the first principal component is significant, indicating that all the variables play a part in the "price" of a player. Further exploration might yield interesting results, for example the third principal component, a function of tournaments won, has a p-value of .0667 indicating the use of the percentage of tournaments won. A further analysis using variable selection (Covered in chapter 6) might be useful.

4. This study relates the amount of evaporation in irrigation (EVAP) with the following variables:
MAXAT: Maximum daily air temperature
MINAT: Minimum daily air temperature
AVAT: A measure of the average air temperature
MAXST: Maximum daily soil temperature
MINST: Minimum daily soil temperature
AVST: A measure of the average soil temperature
MAXH: Maximum daily humidity
MINH: Minimum daily humidity
AVH: A measure of the average daily humidity
WIND: Total wind, measured in miles per day

SAS Code:
The data for this exercise is reg05p04.

```
proc  reg data = reg05p04;
model EVAP = MAXST MINST AVST MAXAT MINAT AVAT MAXH MINH AVH WIND /vif
collin ;
Proc Princomp data = reg05p04 out = prin;
var MAXST MINST AVST MAXAT MINAT AVAT MAXH MINH AVH WIND        ;
;
data reg05p04; set reg05p04; ***irrigation data***;
mast=(maxst+minst)/2;
```

```
rst=Maxst-Minst;
dst=maxst+minst-2*avst;
mat=(maxat+minat)/2;
rat=Maxat-Minat;
dat=maxat+minat-2*avat;
mah=(maxh+minh)/2;
rh=Maxh-Minh;
dh=maxh+minh-2*avh;
proc reg;
model evap = mast mat mah wind;
proc reg;
model evap = rst rat rh wind;
proc reg;
model evap = dst dat dh wind;
proc reg;
model evap = avst avat avh wind;
run;
```

Relevant Output:

The REG Procedure

Dependent Variable: EVAP

Analysis of Variance

Source	DF	Sum of Squares	Mean Square	F Value	Pr > F
Model	11	8190.24226	744.56748	17.44	<.0001
Error	34	1451.86644	42.70195		
Corrected Total	45	9642.10870			

Root MSE	6.53467	R-Square	0.8494	
Dependent Mean	34.67391	Adj R-Sq	0.8007	
Coeff Var	18.84608			

Parameter Estimates

Variable	DF	Parameter Estimate	Standard Error	t Value	Pr > |t|	Variance Inflation
Intercept	1	-54.66839	131.26359	-0.42	0.6797	0
DAY	1	0.12179	0.14432	0.84	0.4046	1.28243
MAXST	1	2.18951	1.00928	2.17	0.0371	39.39147
MINST	1	0.24011	1.10988	0.22	0.8300	14.10252
AVST	1	-0.69828	0.35496	-1.97	0.0574	53.51028
MAXAT	1	0.56646	0.57655	0.98	0.3328	8.99005
MINAT	1	0.14110	0.81536	0.17	0.8636	9.41591
AVAT	1	0.06269	0.22173	0.28	0.7791	22.78171
MAXH	1	1.15557	1.13910	1.01	0.3175	1.98502
MINH	1	0.79857	0.49295	1.62	0.1145	25.70680
AVH	1	-0.56599	0.16268	-3.48	0.0014	24.23633
WIND	1	0.01120	0.00959	1.17	0.2512	2.15561

Collinearity Diagnostics (intercept adjusted)

Number	Eigenvalue	Condition Index	MAXST	MINST	AVST	MAXAT
				----------Proportion of Variation-------------		
1	5.74886	1.00000	0.00072243	0.00174	0.00055192	0.00307
2	2.02857	1.68343	0.00007110	0.00090984	0.00000435	0.00001720
3	1.09974	2.28637	0.00000141	0.00025624	0.00009188	0.00148
4	0.60339	3.08669	0.00008809	0.00520	0.00040532	0.00022769
5	0.23280	4.96937	0.00613	0.10382	0.00414	0.00331
6	0.12055	6.90571	0.00634	0.05769	0.00401	0.54580
7	0.06940	9.10152	0.05591	0.03729	0.01405	0.28276
8	0.06152	9.66679	0.12925	0.14207	0.00036505	0.09973
9	0.02648	14.73533	0.00539	0.42853	0.26066	0.05398
10	0.00871	25.69505	0.79610	0.22250	0.71572	0.00963

Number	MINAT	AVAT	MAXH	MINH	AVH	WIND
1	0.00121	0.00110	0.00067643	0.00043815	0.00071816	0.00003425
2	0.01536	0.00106	0.00015440	0.00505	0.00284	0.06060
3	0.00001435	0.00070614	0.34436	0.00049956	0.00127	0.07386
4	0.00295	0.00048863	0.17550	0.00704	0.00591	0.42911
5	0.07582	0.03846	0.00978	0.00048241	0.02198	0.12944
6	0.11071	0.00111	0.05384	0.00351	0.03817	0.04075
7	0.32580	0.15665	0.03009	0.07954	0.00003368	0.00071127
8	0.25812	0.00940	0.00028379	0.09825	0.13262	0.04090
9	0.07394	0.45265	0.10633	0.15420	0.10163	0.05825
10	0.13607	0.33838	0.27899	0.65098	0.69483	0.16636

Correlation Matrix

	MAXST	MINST	AVST	MAXAT	MINAT	AVAT	MAXH	MINH	AVH	WIND
MAXST	1.0000	0.8506	0.9532	0.9065	0.4656	0.8248	-.1877	-.6748	-.7571	-.0918
MINST	0.8506	1.0000	0.9332	0.8400	0.6809	0.8183	-.1657	-.3398	-.4816	0.0304
AVST	0.9532	0.9332	1.0000	0.9141	0.5941	0.8698	-.1610	-.5331	-.6795	-.0935
MAXAT	0.9065	0.8400	0.9141	1.0000	0.5705	0.8738	-.1034	-.5267	-.6555	-.0899
MINAT	0.4656	0.6809	0.5941	0.5705	1.0000	0.7829	-.1215	0.1909	-.0653	0.4109
AVAT	0.8248	0.8183	0.8698	0.8738	0.7829	1.0000	-.0418	-.3020	-.5366	0.1285
MAXH	-.1877	-.1657	-.1610	-.1034	-.1215	-.0418	1.0000	0.1721	0.2721	-.1469
MINH	-.6748	-.3398	-.5331	-.5267	0.1909	-.3020	0.1721	1.0000	0.9112	0.3457
AVH	-.7571	-.4816	-.6795	-.6555	-.0653	-.5366	0.2721	0.9112	1.0000	0.2229
WIND	-.0918	0.0304	-.0935	-.0899	0.4109	0.1285	-.1469	0.3457	0.2229	1.0000

Eigenvalues of the Correlation Matrix

	Eigenvalue	Difference	Proportion	Cumulative
1	5.74885512	3.72028692	0.5749	0.5749
2	2.02856820	0.92882823	0.2029	0.7777
3	1.09973997	0.49635375	0.1100	0.8877
4	0.60338622	0.37058815	0.0603	0.9481
5	0.23279807	0.11224864	0.0233	0.9713
6	0.12054943	0.05115031	0.0121	0.9834
7	0.06939912	0.00787909	0.0069	0.9903
8	0.06152003	0.03504346	0.0062	0.9965
9	0.02647657	0.01776929	0.0026	0.9991
10	0.00870728		0.0009	1.0000

The PRINCOMP Procedure
Eigenvectors

	Prin1	Prin2	Prin3	Prin4	Prin5
MAXST	0.403973	-.075285	0.007794	0.045702	0.236793
MINST	0.375517	0.161220	0.062995	-.210106	0.583400
AVST	0.407516	0.021495	0.072724	-.113140	0.224596
MAXAT	0.394696	0.017549	0.119722	-.034825	-.082445
MINAT	0.248389	0.526167	0.011842	-.125791	-.396070
AVAT	0.375056	0.218189	0.131347	0.080931	-.445969
MAXH	-.087760	0.024906	0.866046	0.457953	0.067139
MINH	-.252824	0.509710	0.118074	-.328361	-.053384
AVH	-.315534	0.372463	0.183751	-.293223	0.351304
WIND	-.019768	0.493922	-.401515	0.716839	0.244544

Eigenvectors

	Prin6	Prin7	Prin8	Prin9	Prin10
MAXST	0.173262	0.390464	0.558965	0.074891	0.521905
MINST	-.312948	-.190910	-.350827	-.399726	0.165176
AVST	-.159137	0.225912	-.034285	0.601016	-.571122
MAXAT	0.762115	-.416202	-.232729	0.112326	0.027201
MINAT	-.344404	-.448272	0.375668	0.131906	0.102615
AVAT	0.054500	0.491440	-.113370	-.515989	-.255843
MAXH	-.113381	-.064312	-.005880	0.074671	0.069363
MINH	0.103692	0.374273	-.391645	0.321876	0.379263
AVH	0.333105	-.007507	0.443562	-.254733	-.381969
WIND	0.098735	0.009898	-.070664	0.055324	-.053617

Dependent Variable: EVAP

Analysis of Variance

Source	DF	Sum of Squares	Mean Square	F Value	Pr > F
Model	4	6843.76721	1710.94180	25.07	<.0001
Error	41	2798.34148	68.25223		
Corrected Total	45	9642.10870			

Root MSE	8.26149	R-Square	0.7098
Dependent Mean	34.67391	Adj R-Sq	0.6815
Coeff Var	23.82624		

Parameter Estimates

Variable	DF	Parameter Estimate	Standard Error	t Value	Pr > |t|
Intercept	1	38.88843	44.54848	0.87	0.3878
mast	1	-0.53456	0.79103	-0.68	0.5030
mat	1	2.14109	0.78479	2.73	0.0093
mah	1	-1.93956	0.37045	-5.24	<.0001
WIND	1	0.01793	0.00902	1.99	0.0536

Dependent Variable: EVAP

Analysis of Variance

Source	DF	Sum of Squares	Mean Square	F Value	Pr > F
Model	4	6465.45115	1616.36279	20.86	<.0001
Error	41	3176.65754	77.47945		
Corrected Total	45	9642.10870			

Root MSE	8.80224	R-Square	0.6705	
Dependent Mean	34.67391	Adj R-Sq	0.6384	
Coeff Var	25.38577			

Parameter Estimates

Variable	DF	Parameter Estimate	Standard Error	t Value	Pr > |t|
Intercept	1	-32.33763	9.17711	-3.52	0.0011
rst	1	2.34733	0.65245	3.60	0.0009
rat	1	0.29755	0.59421	0.50	0.6192
rh	1	0.32586	0.26026	1.25	0.2177
WIND	1	0.02711	0.01051	2.58	0.0135

Analysis of Variance

Source	DF	Sum of Squares	Mean Square	F Value	Pr > F
Model	4	7824.92562	1956.23141	44.14	<.0001
Error	41	1817.18307	44.32154		
Corrected Total	45	9642.10870			

Root MSE	6.65744	R-Square	0.8115	
Dependent Mean	34.67391	Adj R-Sq	0.7932	
Coeff Var	19.20015			

Parameter Estimates

Variable	DF	Parameter Estimate	Standard Error	t Value	Pr > |t|
Intercept	1	152.03588	24.68632	6.16	<.0001
dst	1	0.09386	0.07148	1.31	0.1965
dat	1	-0.18366	0.06021	-3.05	0.0040
dh	1	0.22054	0.02815	7.84	<.0001
WIND	1	0.01207	0.00748	1.61	0.1143

The REG Procedure

Dependent Variable: EVAP

Analysis of Variance

Source	DF	Sum of Squares	Mean Square	F Value	Pr > F
Model	4	7876.48608	1969.12152	45.73	<.0001
Error	41	1765.62261	43.06397		
Corrected Total	45	9642.10870			

Root MSE	6.56231	R-Square	0.8169
Dependent Mean	34.67391	Adj R-Sq	0.7990
Coeff Var	18.92580		

Parameter Estimates

Variable	DF	Parameter Estimate	Standard Error	t Value	Pr > |t|
Intercept	1	144.74006	27.33022	5.30	<.0001
AVST	1	-0.19621	0.12042	-1.63	0.1109
AVAT	1	0.36497	0.10478	3.48	0.0012
AVH	1	-0.37536	0.04644	-8.08	<.0001
WIND	1	0.01239	0.00740	1.67	0.1016

The output from the regression indicates that the overall model is significant and the relative high R-Square .85 with an adjusted R-Square .80 indicates a reasonable fit. However, only two of the partial regression coefficients are significant. Examination of the diagnostic statistics indicates a severe multicollinearity problem exists.

1 . Determine the extent and nature of multicollinearity.

The variance inflation factor (VIF) for the model is VIF_{model}

$$= \frac{1}{1-R^2} \approx \frac{1}{1-0.8463} \approx 6.506 \,.$$

The VIF's for all but three of the variables exceed the VIF for the model. The pattern might suggest that having three variables to measure air and soil temperature and air and soil humidity may be overkill. The correlation matrix also indicates a strong correlation between the three variables measuring air and soil temperature and humidity.

2. Discuss possible practical reasons why the multicollinearity exists

For each of the following three: daily air temperature, daily soil temperature, and daily humidity, there should be a strong correlation between the minimum value, the maximum value, and the integrated area under their daily curve: a measure of average. This occurs because each of these variables measures the same characteristic.

3. Implement at least one form of remedial action and determine if these methods provide more useful results.

One way to compensate for the multicollinearity would be to use a single variable for each of the following: air temperature, soil temperature, and humidity. Some of these options would be to use the average of minimum and maximum of each, for example: MAST = (MAXST+MINST)/2 would give the average soil temperature; the others could be done similarly. The results are shown in the output and indicate a less than optimum fit. The overall regression is significant but the adjusted R-Square is only 0.68. The only significant partial regression coefficient is the one corresponding to mean humidity and it has a negative value.

A second transformation would be to take the range of the variables, for example RST=MAXST-MINST for soil temperature and similar transformations for air temperature and humidity. These results are shown in the output and indicate a different result. The overall regression is significant but now the adjusted R-Square is only 0.64. The variables rat and wind are significant and both positive.

A third transformation would be look at the shape of the temperature curve or use DST =MAXST + MINST – 2*AVST and similar transformations. These results give a significant regression with an adjusted R-Square of 0.79. The variables dst and dh are significant with dat negative and dh positive.

Finally, since AVST actually measures the integrated area under the daily soil temperature curve, it makes sense to use it as a representative variable for soil temperature and similarly use AVAT and AVH. These results indicate a significant regression, an adjusted R-Square of nearly 0.80, and AVAT positive and significant and AVH negative and significant.

Another way to handle the multicollinearity is to examine the correlational structure of the independent variables. The eigenvalues and the eigenvectors of the correlation matrix suggest that there are several unimportant principal components. The results show that the first four components account for 94.81% of the total variation implying that this set of ten variables essentially has only four factors. This would suggest that an incomplete principal component regression with four components might be useful.

Chapter 6
Problems with the Model
Solutions

1. The gas mileage data for a selection of cars (Chapter 5, Exercise 2) provide a compact data set with reasonably few variables.

SAS Code:
The data for this exercise is reg05p02. We do two variable selection procedures, stepwise and rsquare.

```
proc reg;
model     MPG  =  ESIZE     HP     BARR     WT     TIME/selection =
stepwise;
proc rsquare cp;
model mpg=esize hp barr wt time;
proc reg;
model mpg = wt time/vif;
proc reg;
model mpg = wt hp/vif;
run;
```

Relevant Output:

```
Dependent Variable: MPG
                        Stepwise Selection: Step 1

        Variable WT Entered: R-Square = 0.7528 and C(p) = 11.1979

                        Analysis of Variance
                              Sum of        Mean
        Source          DF    Squares      Square    F Value   Pr > F

        Model            1   847.72525   847.72525     91.38   <.0001
        Error           30   278.32194     9.27740
        Corrected Total 31  1126.04719

                    Parameter    Standard
        Variable    Estimate      Error    Type II SS  F Value  Pr > F

        Intercept    37.28513    1.87763   3658.29412   394.32  <.0001
        WT           -0.00534  0.00055910   847.72525    91.38  <.0001

                Bounds on condition number: 1, 1

                Stepwise Selection: Step 2

        Variable HP Entered: R-Square = 0.8268 and C(p) = 1.4699
```

Analysis of Variance

Source	DF	Sum of Squares	Mean Square	F Value	Pr > F
Model	2	930.99943	465.49972	69.21	<.0001
Error	29	195.04775	6.72578		
Corrected Total	31	1126.04719			

Stepwise Selection: Step 2

Variable	Parameter Estimate	Standard Error	Type II SS	F Value	Pr > F
Intercept	37.22727	1.59879	3646.56391	542.18	<.0001
HP	-0.03177	0.00903	83.27418	12.38	0.0015
WT	-0.00388	0.00063273	252.62656	37.56	<.0001

Bounds on condition number: 1.7666, 7.0665

All variables left in the model are significant at the 0.1500 level.
No other variable met the 0.1500 significance level for entry into the model.

Summary of Stepwise Selection

Step	Variable Entered	Variable Removed	Number Vars In	Partial R-Square	Model R-Square	C(p)	F Value	Pr > F
1	WT		1	0.7528	0.7528	11.1979	91.38	<.0001
2	HP		2	0.0740	0.8268	1.4699	12.38	0.0015

R-Square Selection Method

Number in Model	R-Square	C(p)	Variables in Model
1	0.7528	11.1979	WT
1	0.7183	16.6676	ESIZE
1	0.6024	35.0489	HP
1	0.3035	82.4541	BARR
1	0.1753	102.7887	TIME
2	0.8268	1.4699	HP WT
2	0.8264	1.5284	WT TIME
2	0.7924	6.9163	BARR WT
2	0.7809	8.7419	ESIZE WT
2	0.7737	9.8876	ESIZE BARR
2	0.7482	13.9263	ESIZE HP
2	0.7216	18.1575	ESIZE TIME
2	0.6369	31.5872	HP TIME
2	0.6046	36.6996	HP BARR
2	0.3093	83.5446	BARR TIME
3	0.8348	2.2040	HP WT TIME
3	0.8272	3.3971	BARR WT TIME

3	0.8271	3.4276	HP BARR WT
3	0.8268	3.4618	ESIZE HP WT
3	0.8264	3.5283	ESIZE WT TIME
3	0.8182	4.8350	ESIZE BARR WT
3	0.7837	10.2984	ESIZE BARR TIME
3	0.7740	11.8460	ESIZE HP BARR
3	0.7542	14.9818	ESIZE HP TIME
3	0.6369	33.5859	HP BARR TIME
4	0.8351	4.1443	ESIZE HP WT TIME
4	0.8348	4.2003	HP BARR WT TIME
4	0.8278	5.3097	ESIZE BARR WT TIME
4	0.8276	5.3447	ESIZE HP BARR WT
4	0.7844	12.1853	ESIZE HP BARR TIME
5	0.8361	6.0000	ESIZE HP BARR WT TIME

Analysis of Variance

Source	DF	Sum of Squares	Mean Square	F Value	Pr > F
Model	2	930.99943	465.49972	69.21	<.0001
Error	29	195.04775	6.72578		
Corrected Total	31	1126.04719			

Root MSE	2.59341	R-Square	0.8268
Dependent Mean	20.09063	Adj R-Sq	0.8148
Coeff Var	12.90857		

Parameter Estimates

Variable	DF	Parameter Estimate	Standard Error	t Value	Pr > |t|	Variance Inflation
Intercept	1	37.22727	1.59879	23.28	<.0001	0
WT	1	-0.00388	0.00063273	-6.13	<.0001	1.76662
HP	1	-0.03177	0.00903	-3.52	0.0015	1.76662

Analysis of Variance

Source	DF	Sum of Squares	Mean Square	F Value	Pr > F
Model	2	930.58356	465.29178	69.03	<.0001
Error	29	195.46363	6.74013		
Corrected Total	31	1126.04719			

Root MSE	2.59618	R-Square	0.8264
Dependent Mean	20.09063	Adj R-Sq	0.8144
Coeff Var	12.92232		

Parameter Estimates

	Parameter	Standard				Variance

Variable	DF	Estimate	Error	t Value	Pr > |t|	Inflation
Intercept	1	19.74622	5.25206	3.76	0.0008	0
WT	1	-0.00505	0.00048400	-10.43	<.0001	1.03149
TIME	1	0.92920	0.26502	3.51	0.0015	1.03149

The stepwise procedure identifies a model with two independent variables, HP and WT, as being optimal. The Rsquare procedure also indicates that the best choice is one with two variables. It identifies the same model, but also indicates that a regression using WT and TIME might also be a candidate. Both models are significant and have almost the same R-Square values. However, the one with WT and HP probably is more practical since these are elements easily designed into a vehicle. The "best" model is:

$$MPG = 0.7528WT + 0.0740HP$$

Exercise 2 in Chapter 5 used all the independent variables in a model and found a great deal of multicollinearity. By reducing the number of independent variables to two we have eliminated any multicollinearity (as evidenced by the VIF values). Because of this multicollinearity, however, we must be cautious in using the reduced model.

7. This exercise attempts to estimate the manpower (MANH) needs for operating Bachelor Officer Quarters for the U.S. Navy. The independent variables are:

   OCCUP: Average daily occupancy
   CHECKIN: Monthly average number of check-ins
   HOURS: Weekly hours of service desk operation
   COMMON: Square feet of common-use area
   WINGS: Number of building wings
   CAP: Operational berthing capacity
   ROOMS: Number of rooms.

SAS Code:
The data for this exercise is reg06p07. We first do the regression and diagnostics to determine if there is multicollinearity or if there are outlier problems.

```
proc reg data = reg06p07;
model MANH = OCCUP CHECKIN HOURS COMMON WINGS CAP ROOMS /influence vif;
run;
```

Relevant Output:

Analysis of Variance

Source	DF	Sum of Squares	Mean Square	F Value	Pr > F
Model	7	87387188	12483884	60.26	<.0001
Error	17	3522013	207177		
Corrected Total	24	90909201			

Root MSE	455.16727	R-Square	0.9613		
Dependent Mean	2109.38640	Adj R-Sq	0.9453		
Coeff Var	21.57818				

Parameter Estimates

Variable	DF	Parameter Estimate	Standard Error	t Value	Pr > |t|	Variance Inflation
Intercept	1	134.96790	237.81430	0.57	0.5778	0
OCCUP	1	-1.28377	0.80469	-1.60	0.1291	2.16276
CHECKIN	1	1.80351	0.51624	3.49	0.0028	4.52397
HOURS	1	0.66915	1.84640	0.36	0.7215	1.35735
COMMON	1	-21.42263	10.17160	-2.11	0.0504	2.33264
WINGS	1	5.61923	14.74609	0.38	0.7079	3.65318
CAP	1	-14.48025	4.22018	-3.43	0.0032	37.12912
ROOMS	1	29.32475	6.36590	4.61	0.0003	63.70809

Output Statistics

Obs	Residual	RStudent	Hat Diag H	Cov Ratio	DFFITS
1	-29.7547	-0.0736	0.2573	2.1809	-0.0433
2	-31.1856	-0.0726	0.1609	1.9305	-0.0318
3	-196.1059	-0.4594	0.1614	1.7440	-0.2016
4	-75.5559	-0.1762	0.1631	1.9109	-0.0778
5	-180.7826	-0.4196	0.1475	1.7454	-0.1745
6	-242.9925	-0.5704	0.1589	1.6437	-0.2479
7	313.9226	0.7532	0.1829	1.5041	0.3563
8	-176.0594	-0.4720	0.3591	2.2688	-0.3533
9	128.7439	0.3246	0.2808	2.1428	0.2029
10	39.9941	0.0914	0.1295	1.8581	0.0353
11	635.9230	1.5537	0.1241	0.6025	0.5849
12	184.3228	0.4426	0.2024	1.8475	0.2230
13	-81.7158	-0.1818	0.0802	1.7369	-0.0537
14	-9.9665	-0.0224	0.0969	1.7980	-0.0073
15	-665.2110	-2.5192	0.5576	0.2536	-2.8282
16	-19.6054	-0.0541	0.4024	2.7136	-0.0444
17	-470.9780	-1.3310	0.3682	1.1097	-1.0162
18	418.4622	1.2566	0.4465	1.3820	1.1286
19	437.9938	1.0074	0.0868	1.0874	0.3106
20	-870.0699	-2.8657	0.3663	0.0932	-2.1787
21	826.4958	2.0538	0.0704	0.2687	0.5652
22	-323.8936	-1.6057	0.7854	2.2901	-3.0715
23	-160.2755	-5.2423	0.9885	0.0473	-48.5179
24	413.2654	3.2093	0.8762	0.2461	8.5373
25	135.0289	0.4299	0.5467	3.2687	0.4722

A quick examination of the diffits shows a problem with observation 23. Examination of the data indicates that this record indicates an occupancy of 811.08 and a capacity of only 242. An obvious data entry error, and since we have no additional information we discard that observation. The regression without observation 23 yields:

SAS Code:

```
data reg06p07;
      set reg06p07;
if _n_=23  then delete;
proc reg data = reg06p07;
model MANH = OCCUP CHECKIN HOURS COMMON WINGS CAP ROOMS /influence vif;
run;
```

Relevant Output:

Analysis of Variance

Source	DF	Sum of Squares	Mean Square	F Value	Pr > F
Model	7	87497673	12499668	154.32	<.0001
Error	16	1295987	80999		
Corrected Total	23	88793659			

Root MSE		284.60353	R-Square	0.9854	
Dependent Mean		2050.00708	Adj R-Sq	0.9790	
Coeff Var		13.88305			

Parameter Estimates

Variable	DF	Parameter Estimate	Standard Error	t Value	Pr > |t|	Variance Inflation
Intercept	1	171.47336	148.86168	1.15	0.2663	0
OCCUP	1	21.04562	4.28905	4.91	0.0002	43.63222
CHECKIN	1	1.42632	0.33071	4.31	0.0005	4.54154
HOURS	1	-0.08927	1.16353	-0.08	0.9398	1.36076
COMMON	1	7.65033	8.43835	0.91	0.3781	4.06083
WINGS	1	-5.30231	9.45276	-0.56	0.5826	3.79996
CAP	1	-4.07475	3.30195	-1.23	0.2350	56.60333
ROOMS	1	0.33191	6.81399	0.05	0.9618	178.70159

Notice that the multicollinearity is still a problem, especially concerning OCCUP, CAP, and ROOMS. Since bigger BOQ's will have more of everything, an appropriate redefinition to a per room basis might correct the multicollinearity. This is done using the following SAS statements:

SAS Code:

```
data reg06p07;
      set reg06p07; *** BOQ data ***;
ROCCUP=OCCUP/ROOMS;
RCHECKIN=CHECKIN/ROOMS;
RCOMMON=COMMON/ROOMS;
RWINGS=WINGS/ROOMS;
RCAP=CAP/ROOMS;
RMANH=MANH/ROOMS;
proc reg data = reg06p07;
model RMANH =ROCCUP RCHECKIN HOURS RCOMMON RWINGS RCAP ROOMS / vif;
run;
```

Relevant Output:

Analysis of Variance

Source	DF	Sum of Squares	Mean Square	F Value	Pr > F
Model	7	775.01435	110.71634	1.76	0.1659
Error	16	1008.44749	63.02797		
Corrected Total	23	1783.46183			

Root MSE	7.93902	R-Square	0.4346	
Dependent Mean	19.96950	Adj R-Sq	0.1872	
Coeff Var	39.75571			

Parameter Estimates

Variable	DF	Parameter Estimate	Standard Error	t Value	Pr > |t|	Variance Inflation
Intercept	1	36.88142	21.07922	1.75	0.0993	0
ROCCUP	1	1.66891	10.01957	0.17	0.8698	2.75373
RCHECKIN	1	1.58059	1.08598	1.46	0.1649	2.09438
HOURS	1	-0.02174	0.03670	-0.59	0.5619	1.73989
RCOMMON	1	-12.94336	25.50015	-0.51	0.6187	1.68052
RWINGS	1	20.20721	26.22190	0.77	0.4522	1.22610
RCAP	1	-16.02822	13.56252	-1.18	0.2546	1.52130
ROOMS	1	-0.01943	0.01818	-1.07	0.3010	1.63477

The multicollinearity has disappeared, however, the model is no longer significant. In other words the usefulness of the model is questionable. A stepwise variable selection yields:

Relevant Output:

Stepwise Selection: Step 1

Variable RCAP Entered: R-Square = 0.1849 and C(p) = 3.0633

Analysis of Variance

Source	DF	Sum of Squares	Mean Square	F Value	Pr > F
Model	1	329.82588	329.82588	4.99	0.0360
Error	22	1453.63595	66.07436		
Corrected Total	23	1783.46183			

Variable	Parameter Estimate	Standard Error	Type II SS	F Value	Pr > F

Intercept	46.25587	11.88177	1001.39207	15.16	0.0008
RCAP	-25.15406	11.25854	329.82588	4.99	0.0360

Bounds on condition number: 1, 1

Stepwise Selection: Step 2

Variable RCHECKIN Entered: R-Square = 0.2908 and C(p) = 2.0666

Analysis of Variance

Source	DF	Sum of Squares	Mean Square	F Value	Pr > F
Model	2	518.70510	259.35255	4.31	0.0271
Error	21	1264.75674	60.22651		
Corrected Total	23	1783.46183			

Stepwise Selection: Step 2

Variable	Parameter Estimate	Standard Error	Type II SS	F Value	Pr > F
Intercept	44.18015	11.40420	903.88394	15.01	0.0009
RCHECKIN	1.30004	0.73411	188.87921	3.14	0.0911
RCAP	-25.90676	10.75719	349.31401	5.80	0.0253

Bounds on condition number: 1.0016, 4.0063

Stepwise Selection: Step 3

Variable ROOMS Entered: R-Square = 0.3868 and C(p) = 1.3521

Analysis of Variance

Source	DF	Sum of Squares	Mean Square	F Value	Pr > F
Model	3	689.79536	229.93179	4.20	0.0185
Error	20	1093.66647	54.68332		
Corrected Total	23	1783.46183			

Variable	Parameter Estimate	Standard Error	Type II SS	F Value	Pr > F
Intercept	39.37033	11.20177	675.49121	12.35	0.0022
RCHECKIN	1.53970	0.71251	255.35652	4.67	0.0430
RCAP	-18.83125	11.00307	160.17146	2.93	0.1025
ROOMS	-0.02562	0.01448	171.09027	3.13	0.0922

Bounds on condition number: 1.1961, 10.168

All variables left in the model are significant at the 0.1500 level.
No other variable met the 0.1500 significance level for entry into the model.

A significant regression equation does emerge with the independent variables RCHECKIN and ROOMS significant, but with an R-square of 0.30. Another significant model emerges with the variables RCHECKIN, RCAP, and ROOMS. However, at the 0.05 level, only RCHECKIN is significant, and the R-square is only 0.39. Realistically, the only variable that appears to affect MANH is the number of Rooms.

Another way to approach this problem is to choose a subset of independent variables without the transformations. Stepwise, R-square and Cp statistic were used to select the variables. All three procedures identify the three-variable model which includes occup, checkin and cap. The R-square value for this model is 0.9851. Furthermore, by reducing the number of independent variables, the problem of multicollinearity is eliminated. The individual VIF values are all small the $VIF_{model} = 67.11$. The "best" model is:

$$MANH = 207.865 + 20.67OCCUP + 1.44CHECKIN - 3.45CAP$$

SAS Code:

```
proc reg;
id id;
model MANH = OCCUP CHECKIN HOURS COMMON WINGS CAP ROOMS /influence vif;
proc reg;
id id;
model MANH = OCCUP CHECKIN HOURS COMMON WINGS CAP ROOMS
/selection=stepwise;
proc rsquare cp;
model MANH = OCCUP CHECKIN HOURS COMMON WINGS CAP ROOMS;
proc reg;
model MANH = OCCUP CHECKIN CAP /vif;
run;
```

Relevant Output:

Stepwise Selection: Step 1

Variable OCCUP Entered: R-Square = 0.9619 and C(p) = 21.7606

Analysis of Variance

Source	DF	Sum of Squares	Mean Square	F Value	Pr > F
Model	1	85411083	85411083	555.51	<.0001
Error	22	3382576	153753		
Corrected Total	23	88793659			

Variable	Parameter Estimate	Standard Error	Type II SS	F Value	Pr > F
Intercept	163.06377	113.20758	318999	2.07	0.1638
OCCUP	21.08503	0.89460	85411083	555.51	<.0001

Bounds on condition number: 1, 1

Stepwise Selection: Step 2
Variable CHECKIN Entered: R-Square = 0.9765 and C(p) = 7.8089

Analysis of Variance

Source	DF	Sum of Squares	Mean Square	F Value	Pr > F
Model	2	86703162	43351581	435.49	<.0001
Error	21	2090497	99547		
Corrected Total	23	88793659			

Stepwise Selection: Step 2

Variable	Parameter Estimate	Standard Error	Type II SS	F Value	Pr > F
Intercept	170.79057	91.11681	349753	3.51	0.0749
OCCUP	16.77062	1.39724	14341264	144.06	<.0001
CHECKIN	1.20306	0.33393	1292079	12.98	0.0017

Bounds on condition number: 3.7677, 15.071

Stepwise Selection: Step 3
Variable CAP Entered: R-Square = 0.9839 and C(p) = 1.7003

Analysis of Variance

Source	DF	Sum of Squares	Mean Square	F Value	Pr > F
Model	3	87359949	29119983	406.22	<.0001
Error	20	1433710	71686		
Corrected Total	23	88793659			

Variable	Parameter Estimate	Standard Error	Type II SS	F Value	Pr > F
Intercept	207.86486	78.28539	505396	7.05	0.0152
OCCUP	20.67163	1.75123	9988307	139.34	<.0001
CHECKIN	1.43624	0.29366	1714758	23.92	<.0001
CAP	-3.45397	1.14110	656787	9.16	0.0067

Bounds on condition number: 8.2191, 59.71

All variables left in the model are significant at the 0.1500 level.

No other variable met the 0.1500 significance level for entry into the model.

Summary of Stepwise Selection

Step	Variable Entered	Variable Removed	Number Vars In	Partial R-Square	Model R-Square	C(p)	F Value	Pr > F
1	OCCUP		1	0.9619	0.9619	21.7606	555.51	<.0001
2	CHECKIN		2	0.0146	0.9765	7.8089	12.98	0.0017
3	CAP		3	0.0074	0.9839	1.7003	9.16	0.0067

R-Square Selection Method

Number in Model	R-Square	C(p)	Variables in Model
1	0.9619	21.7606	OCCUP
1	0.8888	101.8743	ROOMS
1	0.8149	182.8633	CHECKIN
1	0.7920	207.9875	CAP
1	0.5362	488.4304	WINGS
1	0.3196	725.8567	COMMON
1	0.2321	821.8176	HOURS
2	0.9765	7.8089	OCCUP CHECKIN
2	0.9645	20.8704	OCCUP CAP
2	0.9634	22.1085	OCCUP ROOMS
2	0.9629	22.7211	OCCUP WINGS
2	0.9621	23.5391	OCCUP HOURS
2	0.9620	23.6700	OCCUP COMMON
2	0.9238	65.5207	CHECKIN ROOMS
2	0.9136	76.6623	CAP ROOMS
2	0.8966	95.3875	CHECKIN WINGS
2	0.8941	98.0390	COMMON ROOMS
2	0.8899	102.6530	HOURS ROOMS
2	0.8895	103.0853	WINGS ROOMS
2	0.8714	123.0140	CHECKIN CAP
2	0.8423	154.8833	CHECKIN COMMON
2	0.8248	174.0488	WINGS CAP
2	0.8205	178.8267	CHECKIN HOURS
2	0.7981	203.2758	HOURS CAP
2	0.7948	206.9972	COMMON CAP
2	0.5866	435.2032	HOURS WINGS
2	0.5442	481.6478	COMMON WINGS
2	0.4026	636.8476	HOURS COMMON
3	0.9839	1.7003	OCCUP CHECKIN CAP
3	0.9803	5.6299	OCCUP CHECKIN ROOMS
3	0.9767	9.5345	OCCUP CHECKIN COMMON
3	0.9766	9.6912	OCCUP CHECKIN WINGS
3	0.9765	9.8083	OCCUP CHECKIN HOURS
3	0.9661	21.1552	OCCUP WINGS CAP
3	0.9649	22.4518	OCCUP COMMON CAP
3	0.9648	22.5407	OCCUP HOURS CAP
3	0.9647	22.7061	OCCUP CAP ROOMS
3	0.9647	22.7347	OCCUP COMMON ROOMS

3	0.9642	23.2838	OCCUP WINGS ROOMS
3	0.9637	23.7877	OCCUP HOURS ROOMS
3	0.9636	23.9483	OCCUP COMMON WINGS
3	0.9631	24.4881	OCCUP HOURS WINGS
3	0.9622	25.4849	OCCUP HOURS COMMON
3	0.9541	34.2710	CHECKIN CAP ROOMS
3	0.9330	57.4604	COMMON CAP ROOMS
3	0.9314	59.2114	CHECKIN WINGS ROOMS
3	0.9245	66.7297	CHECKIN COMMON ROOMS
3	0.9240	67.3442	CHECKIN HOURS ROOMS
3	0.9177	74.1786	WINGS CAP ROOMS
3	0.9144	77.8024	HOURS CAP ROOMS
3	0.9102	82.4206	CHECKIN WINGS CAP
3	0.8976	96.2992	COMMON WINGS ROOMS
3	0.8973	96.5463	CHECKIN HOURS WINGS
3	0.8966	97.3875	CHECKIN COMMON WINGS
3	0.8958	98.2169	HOURS COMMON ROOMS
3	0.8906	103.8966	HOURS WINGS ROOMS
3	0.8768	119.0394	CHECKIN COMMON CAP
3	0.8728	123.4640	CHECKIN HOURS CAP
3	0.8439	155.1323	CHECKIN HOURS COMMON
3	0.8286	171.9173	HOURS WINGS CAP
3	0.8264	174.3306	COMMON WINGS CAP
3	0.7998	203.4459	HOURS COMMON CAP
3	0.5889	434.6054	HOURS COMMON WINGS
4	0.9851	2.3204	OCCUP CHECKIN COMMON CAP
4	0.9845	2.9396	OCCUP CHECKIN CAP ROOMS
4	0.9839	3.6414	OCCUP CHECKIN COMMON ROOMS
4	0.9839	3.6868	OCCUP CHECKIN WINGS CAP
4	0.9839	3.6869	OCCUP CHECKIN HOURS CAP
4	0.9807	7.1615	OCCUP CHECKIN WINGS ROOMS
4	0.9803	7.6117	OCCUP CHECKIN HOURS ROOMS
4	0.9767	11.5233	OCCUP CHECKIN COMMON WINGS
4	0.9767	11.5310	OCCUP CHECKIN HOURS COMMON
4	0.9766	11.6912	OCCUP CHECKIN HOURS WINGS
4	0.9683	20.7775	OCCUP COMMON WINGS CAP
4	0.9674	21.7720	OCCUP COMMON WINGS ROOMS
4	0.9674	21.7853	OCCUP WINGS CAP ROOMS
4	0.9664	22.7917	OCCUP HOURS WINGS CAP
4	0.9651	24.2143	OCCUP HOURS COMMON CAP
4	0.9650	24.3954	OCCUP HOURS CAP ROOMS
4	0.9649	24.4337	OCCUP COMMON CAP ROOMS
4	0.9648	24.5624	OCCUP HOURS COMMON ROOMS
4	0.9645	24.9575	OCCUP HOURS WINGS ROOMS
4	0.9637	25.8305	OCCUP HOURS COMMON WINGS
4	0.9630	26.5730	CHECKIN COMMON CAP ROOMS
4	0.9542	36.2435	CHECKIN HOURS CAP ROOMS
4	0.9542	36.2567	CHECKIN WINGS CAP ROOMS
4	0.9348	57.4969	CHECKIN COMMON WINGS ROOMS
4	0.9347	57.6132	HOURS COMMON CAP ROOMS
4	0.9340	58.3315	COMMON WINGS CAP ROOMS
4	0.9315	61.1399	CHECKIN HOURS WINGS ROOMS
4	0.9248	68.4362	CHECKIN HOURS COMMON ROOMS
4	0.9185	75.3220	HOURS WINGS CAP ROOMS
4	0.9108	83.8280	CHECKIN COMMON WINGS CAP
4	0.9105	84.0967	CHECKIN HOURS WINGS CAP

4	0.8993	96.3615	HOURS COMMON WINGS ROOMS
4	0.8973	98.5287	CHECKIN HOURS COMMON WINGS
4	0.8775	120.3097	CHECKIN HOURS COMMON CAP
4	0.8308	171.4557	HOURS COMMON WINGS CAP
5	0.9854	4.0091	OCCUP CHECKIN COMMON WINGS CAP
5	0.9851	4.3159	OCCUP CHECKIN HOURS COMMON CAP
5	0.9851	4.3195	OCCUP CHECKIN COMMON CAP ROOMS
5	0.9846	4.8286	OCCUP CHECKIN WINGS CAP ROOMS
5	0.9846	4.9347	OCCUP CHECKIN HOURS CAP ROOMS
5	0.9840	5.5567	OCCUP CHECKIN COMMON WINGS ROOMS
5	0.9839	5.6109	OCCUP CHECKIN HOURS COMMON ROOMS
5	0.9839	5.6747	OCCUP CHECKIN HOURS WINGS CAP
5	0.9807	9.1505	OCCUP CHECKIN HOURS WINGS ROOMS
5	0.9767	13.5201	OCCUP CHECKIN HOURS COMMON WINGS
5	0.9684	22.6387	OCCUP HOURS COMMON WINGS CAP
5	0.9683	22.7526	OCCUP COMMON WINGS CAP ROOMS
5	0.9676	23.4710	OCCUP HOURS WINGS CAP ROOMS
5	0.9674	23.6852	OCCUP HOURS COMMON WINGS ROOMS
5	0.9651	26.2070	OCCUP HOURS COMMON CAP ROOMS
5	0.9633	28.2636	CHECKIN HOURS COMMON CAP ROOMS
5	0.9632	28.3797	CHECKIN COMMON WINGS CAP ROOMS
5	0.9542	38.2277	CHECKIN HOURS WINGS CAP ROOMS
5	0.9356	58.5503	HOURS COMMON WINGS CAP ROOMS
5	0.9350	59.2427	CHECKIN HOURS COMMON WINGS ROOMS
5	0.9112	85.3707	CHECKIN HOURS COMMON WINGS CAP
6	0.9854	6.0024	OCCUP CHECKIN HOURS COMMON WINGS CAP
6	0.9854	6.0059	OCCUP CHECKIN COMMON WINGS CAP ROOMS
6	0.9851	6.3146	OCCUP CHECKIN HOURS COMMON CAP ROOMS
6	0.9847	6.8219	OCCUP CHECKIN HOURS WINGS CAP ROOMS
6	0.9840	7.5229	OCCUP CHECKIN HOURS COMMON WINGS ROOMS
6	0.9684	24.6012	OCCUP HOURS COMMON WINGS CAP ROOMS
6	0.9634	30.0769	CHECKIN HOURS COMMON WINGS CAP ROOMS
7	0.9854	8.0000	OCCUP CHECKIN HOURS COMMON WINGS CAP ROOMS

Dependent Variable: MANH
Analysis of Variance

Source	DF	Sum of Squares	Mean Square	F Value	Pr > F
Model	3	87359949	29119983	406.22	<.0001
Error	20	1433710	71686		
Corrected Total	23	88793659			

Root MSE	267.74149	R-Square	0.9839
Dependent Mean	2050.00708	Adj R-Sq	0.9814
Coeff Var	13.06052		

Parameter Estimates

Variable	DF	Parameter Estimate	Standard Error	t Value	Pr > |t|	Variance Inflation
Intercept	1	207.86486	78.28539	2.66	0.0152	0
OCCUP	1	20.67163	1.75123	11.80	<.0001	8.21908
CHECKIN	1	1.43624	0.29366	4.89	<.0001	4.04615
CAP	1	-3.45397	1.14110	-3.03	0.0067	7.63825

Chapter 7
Curve Fitting
Solutions

2.This exercise concerns an experiment on effect of certain soil nutrients on the yield of ryegrass. The experimental unit is a pot with 20 ryegrass plants and the response variable is dry matter (YIELD) in grams. The nutrients in this study are calcium (CA), aluminum (AL), and phosphorus (P) in parts per million. The experimental design is a composite design with eight replications at the center points. Fit a quadratic response surface and produce plots to interpret results.

SAS Code:
The data for this exercise is reg07p02.
First, we perform the response surface analysis and search for an optimum response.

```
proc rsreg data = reg07p02;
model yield =ca al p/lackfit ;
ridge min;
run;

data p1 ;
    flag=1;
        do ca = 0 to 400 by 40;
        do al = 0 to 100 by 10;
                do p = 0 to 80 by 8;
                    output;
                end;
            end;
        end;
data p2 ;
    set reg07p02 p1;
proc rsreg data = p2 out = p3;
    model yield = ca al p /predict residual ;
    id flag;
;
data plot (rename  = (yield = predict))
        p4 (rename = (yield = predict))
        p5 (rename = (yield = residual));
set p3;
    If flag = 1 and _type_ = 'PREDICT' THEN OUTPUT plot;
    If flag = . and _type_ = 'PREDICT' THEN OUTPUT p4;
    If flag = . and _type_ = 'RESIDUAL' THEN OUTPUT p5;
data p6;
    merge p4 p5;
run;
proc gplot data = p6;
plot residual*PREDICT /VREF = 0;
PROC G3D DATA= plot;
PLOT al*p = predict /zmin = -3 zmax = 3 ;
by ca;
run;
```

Relevant Output:

The RSREG Procedure

Response Surface for Variable yield

(1)Response Mean	0.891032
Root MSE	0.165695
R-Square	0.9576
Coefficient of Variation	18.5959

(2)Regression	DF	Type I Sum of Squares	R-Square	F Value	Pr > F
Linear	3	6.478334	0.8333	78.65	<.0001
Quadratic	3	0.802963	0.1033	9.75	0.0015
Crossproduct	3	0.163314	0.0210	1.98	0.1704
Total Model	9	7.444611	0.9576	30.13	<.0001

(3)Residual	DF	Sum of Squares	Mean Square	F Value	Pr > F
Lack of Fit	5	0.021411	0.004282	0.10	0.9896
Pure Error	7	0.308048	0.044007		
Total Error	12	0.329460	0.027455		

Parameter Estimate (4)

Parameter	DF	Estimate	Standard Error	t Value	Pr > |t|	from Coded Data
Intercept	1	3.453683	0.693188	4.98	0.0003	0.738558
ca	1	-0.010406	0.002937	-3.54	0.0040	-0.559140
al	1	-0.060990	0.011747	-5.19	0.0002	-1.135945
p	1	0.019637	0.014684	1.34	0.2059	0.609994
ca*ca	1	0.000008037	0.000003272	2.46	0.0302	0.321463
al*ca	1	0.000085016	0.000036614	2.32	0.0386	0.850156
al*al	1	0.000267	0.000052350	5.10	0.0003	0.667563
p*ca	1	0.000003613	0.000045767	0.08	0.9384	0.028906
p*al	1	-0.000136	0.000183	-0.74	0.4723	-0.271719
p*p	1	0.000021040	0.000081796	0.26	0.8014	0.033663

The RSREG Procedure

(5) Factor	DF	Sum of Squares	Mean Square	F Value	Pr > F
ca	4	1.339290	0.334822	12.20	0.0003
al	4	5.109890	1.277473	46.53	<.0001
p	4	1.237572	0.309393	11.27	0.0005

```
                      The RSREG Procedure
       (6)Canonical Analysis of Response Surface Based on Coded Data

                                 Critical Value
                 Factor        Coded        Uncoded

                   ca        -1.884014    -176.802784
                   al         2.077889     153.894447
                   p          0.134675      45.387017

               Predicted value at stationary point: 0.126164
                      Stationary point is a saddle point.
```

To interpret the results, we will answer the following questions:

Is the model adequate?
This can only be answered if there are replications; in this case we have eight replicated observations at the center of the design providing a seven degree-of-freedom estimate of the pure error. The lack of fit test (3) shows that the p-value of the lack of fit is 0.9896. Thus, the model is very adequate.
Do we need all three factors?
The question is answered by (5) of the output. Here, we are given the tests for the elimination of all terms involving each factor. In other words, the test for ca is the test for the deletion of ca, ca^2, ca *al, and ca * p, leaving a model with only the other two factors. In this output, we can see that all three ca, al, and p are very important.
Do we need quadratic and cross-product terms?
This portion (2) gives sequential (SAS Type I) sum of squares for using first only the three linear terms. It then adds the three quadratics, and finally the three cross-product terms. Here, we see that both the linear and quadratics terms are definitely needed, but the product terms are not. The last line shows that the overall model is definitely significant.
In addition portion (1) of the output gives some overall statistics and portion (4) gives the coefficients and their statistics. Here, we can see that one product term (p x ca) is indeed significant at the 0.05 level; hence a decision to omit all product terms would be a mistake.

Often response surface experiments are performed to find some optimum level of the response. In this application, for example, we would like to see what levels of the factors produce the minimum amount of dry matter (YIELD). Portion (6) of the output attempts to answer the question. The first statistics identify the "critical values", which give the factor levels and estimated response at a "stationary point", that is, a point in which the response has no slope. Now, by laws of geometry, there is for a quadratic response function only one stationary point, which can be a maximum, minimum, or saddle point.

Although the experiment did not provide us with the desired minimum response, it is of interest to examine the nature of the response surface. This is somewhat difficult for the three-factor experiment. Thus, we need to examine all two-factor response curves for various levels of the third. In this example, we plot the response curve for p by al for levels of ca = 0, 200 and 400. The SAS code and plots are shown below.

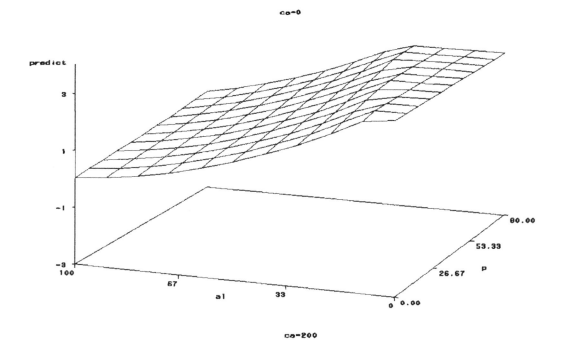

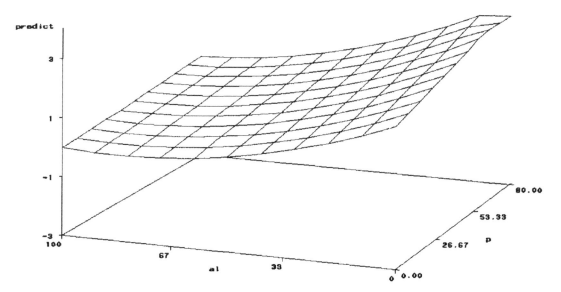

68

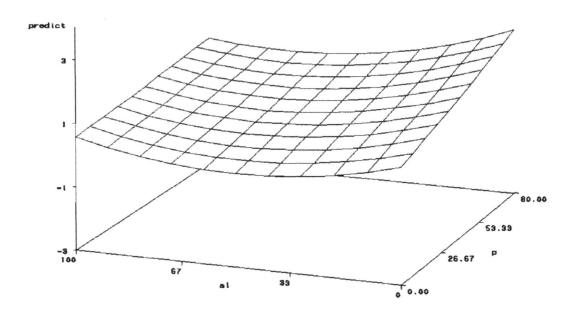

The response curves indicate a nearly linear upward response to p and a concave, slightly positive response to al. For the different levels of ca, the upward response to al decreases and the slight downward curve at the higher levels of ca and p disappears. Graphically, it is very hard to choose the levels of al, p, and ca that would minimize the response.

7. Removing iron from vegetable oil increases its shelf life. One method of reducing the amount of iron is to add some phosphoric acid in a water solution; iron will then precipitate out. The data in table 7.18 gives the amount of iron (dependent variable) remaining after adding various amounts of phosphoric acid (PHOS) in a fixed amount of water. Fit a polynomial curve and determine the amount of phosphoric acid that will give the maximum precipitation of the iron.

SAS Code:
The data for this exercise is reg07p07.
SAS code for data input and creating powers of independent variable.

```
data reg07p07;
input PHOS    IRON  ;
cards;
0.05    0.33
0.10    0.19
0.15    0.10
0.20    0.25
0.25    0.17
0.30    0.12
0.35    0.12
0.40    0.12
0.50    0.12
```

```
0.60      0.12
0.80      0.12
0.90      0.07
1.00      0.18
1.50      0.14
2.00      0.17
;
data powers; set reg07p07;
phos2 = phos**2;
phos3 = phos**3;
phos4 = phos**4;
phos5 = phos**5;
```

We now use PROC GLM requesting sequential sums of squares to determine appropriate degree of equation.

```
proc glm;
model iron=phos phos2 phos3 phos4 phos5/ss1;
run;
```

Relevant Output:

```
                    The GLM Procedure
```

Dependent Variable: IRON

Source	DF	Sum of Squares	Mean Square	F Value	Pr > F
Model	5	0.03285198	0.00657040	2.20	0.1440
Error	9	0.02692136	0.00299126		
Corrected Total	14	0.05977333			

R-Square	Coeff Var	Root MSE	IRON Mean
0.549609	35.36148	0.054692	0.154667

Source	DF	Type I SS	Mean Square	F Value	Pr > F
PHOS	1	0.00320943	0.00320943	1.07	0.3273
phos2	1	0.01678326	0.01678326	5.61	0.0420
phos3	1	0.00848327	0.00848327	2.84	0.265
phos4	1	0.00391700	0.00391700	1.31	0.2820
phos5	1	0.00045901	0.00045901	0.15	0.7044

Parameter	Estimate	Standard Error	t Value	Pr > |t|
Intercept	0.342163691	0.08206499	4.17	0.0024
PHOS	-1.362747428	0.98888289	-1.38	0.2015
phos2	3.001454651	3.60754193	0.83	0.4269
phos3	-3.031541591	5.17845916	-0.59	0.5727
phos4	1.431996527	3.09924505	0.46	0.6550
phos5	-0.253516834	0.64717412	-0.39	0.7044

Actually the regression is not significant, but the sequential sums of squares (should be used to check if the higher order term(s) are necessary) suggest the quadratic term may be useful but the third, fourth, and fifth degree terms are not. We now use PROC REG to fit the quadratic polynomial and plot the results.

SAS Code:

```
proc reg;
model iron=phos phos2;
output out= resid
p=prediron r=resiron;
symbol1 v=star c=black;
symbol2 v=none l=1 i=spline c=black;
proc gplot;
plot iron*phos=1 prediron*phos=2 / overlay;
plot resiron*phos=1 / vref=0;
run;
```

Relevant Output:

The REG Procedure

Dependent Variable: IRON

Analysis of Variance

Source	DF	Sum of Squares	Mean Square	F Value	Pr > F
Model	2	0.01999	0.01000	3.02	0.0869
Error	12	0.03978	0.00332		
Corrected Total	14	0.05977			

Root MSE	0.05758	R-Square	0.3345	
Dependent Mean	0.15467	Adj R-Sq	0.2236	
Coeff Var	37.22618			

Parameter Estimates

Variable	DF	Parameter Estimate	Standard Error	t Value	Pr > |t|
Intercept	1	0.22465	0.03268	6.87	<.0001
PHOS	1	-0.23134	0.09483	-2.44	0.0312
phos2	1	0.10747	0.04776	2.25	0.0440

The regression is still not significant, but the quadratic term is. The plot of the predicted curve is the following:

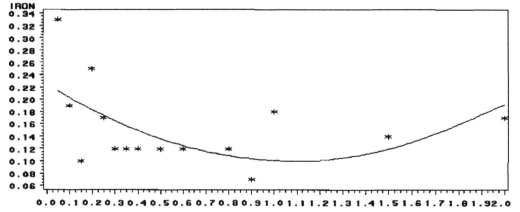

This is a typica

l quadratic with a minimum.

The residual plot.

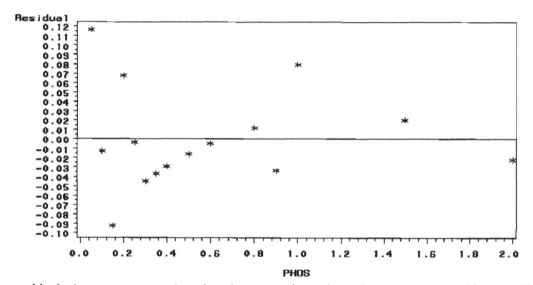

The residuals do not appear random, but the pattern is too irregular to suggest anything specific.

Estimating the point of minimum predicted iron requires the first derivative, which results in:
$$-.0231 + 0.214(PHOS) = 0,$$
which results in the minimum at PHOS = 1.08, where the estimated IRON is 0.100.

Chapter 8
Introduction to
Nonlinear Models
Solutions

1. Exercise 1 of Chapter 7 examined the effect of drought conditions on the weight of pine seedlings. Six beds containing 24 seedlings were subjected to drought over 12 days. The response is the average weight (WEIGHT) in each bed, the dependent variable is time in DAYS. Use a nonlinear regression to see if the decay models are appropriate for these data.

SAS Code:
The data for this exercise is reg07p01.

```
proc nlin data = reg07p01;
     parms b = 105 c = -0.2036;
     model weight = b*exp(c*day);
proc univariate ;
     var Weight;
proc nlin data = reg07p01;
     parms  a = 0 b = 105 c = -0.2036 ;
     model weight = a+b*exp(c*day);
proc univariate ;
     var Weight;

run;
```

Relevant Output:

starting values b = 105 c= -0.2036

model : weight = b*exp(c*day);

The NLIN Procedure
Dependent Variable WEIGHT
Method: Gauss-Newton

Iterative Phase

Iter	b	c	Sum of Squares
0	105.0	-0.1400	18084.5
1	81.9887	-0.0697	6603.0
2	81.0387	-0.0645	6503.1
3	80.8408	-0.0639	6502.0
4	80.8173	-0.0638	6502.0
5	80.8144	-0.0638	6502.0
6	80.8140	-0.0638	6502.0

NOTE: An intercept was not specified for this model.

Source	DF	Sum of Squares	Mean Square	F Value	Approx Pr > F
Model	2	256372	128186	1380.04	<.0001
Error	70	6502.0	92.8856		
Uncorrected Total	72	262874			

The NLIN Procedure

Parameter	Estimate	Approx Std Error	Approximate 95% Confidence Limits	
b	80.8139	2.4479	75.9318	85.6961
c	-0.0638	0.00584	-0.0754	-0.0522

Model with Intercept
a = 0 b= 105 c= -0.2036
weight = a+b*exp(c*day);

The NLIN Procedure
Dependent Variable WEIGHT
Method: Gauss-Newton

Iterative Phase

Iter	a	b	c	Sum of Squares
0	0	105.0	-0.2036	37939.9
1	69.0826	28.6981	-0.3613	28852.1
2	53.4970	47.7256	-0.8415	670.2
3	50.6893	53.6697	-0.8605	169.4
4	50.7032	53.6854	-0.8617	169.4
5	50.7041	53.6862	-0.8618	169.4
6	50.7042	53.6863	-0.8619	169.4
7	50.7043	53.6863	-0.8619	169.4

Source	DF	Sum of Squares	Mean Square	F Value	Approx Pr > F
Model	2	16729.4	8364.7	3407.49	<.0001
Error	69	169.4	2.4548		
Corrected Total	71	16898.8			

The NLIN Procedure

Parameter	Estimate	Approx Std Error	Approximate 95% Confidence Limits	
a	50.7043	0.2404	50.2247	51.1839
b	53.6863	0.6597	52.3703	55.0023
c	-0.8619	0.0244	-0.9105	-0.8132

The objective is to fit one of the decay models used in Example 8.4. The first of which is:

$$weight = \beta_0 e^{-\beta_1 day} + \varepsilon ,$$

where β_0 is the initial count at day 0, and β_1 is the exponential decay rate. The ε are assumed to be independent, normal errors. The model is nonlinear since the error is not multiplicative.

In order to do the nonlinear regression, we need starting values for the iterative process that will estimate the coefficients. To get a starting value for β_0, we observe that the first piece of data has a value of day = 0, giving us a good estimate for β_0 of 105. To get a starting value for β_1, we choose the day 3 where weight = 57. This gives us the following equation:

$$57 = 105 e^{-\beta_1 (3)}$$

Taking natural logarithms and solving gives us an estimate of $\beta_1 = 0.2036$ (the negative sign is part of the model; hence, the exponent is -0.2036.
We will use these starting values in the PROC NLIN nonlinear regression option in SAS. This procedure produces estimates with the minimum value of the residual sums of squares. The estimates of the parameters are $\widehat{\beta_0} = 80.8139$ and $\widehat{\beta_1} = 0.0638$.
The overall model is significant as indicated by the F test. Furthermore, we can be approximately 95% confident that the true value of β_0 is between 75.9 and 85.7, and the true value of β_1 is between 0.052 and 0.075. Since neither confidence interval contains 0, we conclude that both coefficients are significantly different from 0.

The estimated coefficients can be used to provide an estimated model:

$$\widehat{\mu}_{weight/day} = 80.8 e^{-0.064 day}$$

The second model used in Example 8..4 has three parameters and is of the form:

$$weight = \beta_0 + \beta_1 e^{-\beta_2 day} + \varepsilon$$

This decay model is also significant as indicated by the F test (the p-value of the model is less than 0.0001). The estimated decay model has the form:

$$weight = 50.7043 + 53.6863 e^{-0.8619 day}$$

We note that the asymptotic standard error of the decay coefficient is significantly higher in three parameter model (0.0244) than in the two parameters model (0.00584). The partitioning sum of squares for the three parameters model shows that model fits better (much less SSE) for three parameter model than the two parameter one.

2. The data in Table 8.21 resulted from a kinetics study in which the velocity of a reaction (y) was expected to be related to the concentration (x) with the equation

$$y = \frac{\beta_0 x}{\beta_1 + x} + \varepsilon$$

SAS Code:
The data for this exercise is reg08p03.

```
data reg08p03new ;set reg08p03;
z= 1/veloc ;
w = 1/conc;
;
proc reg data=  reg08p03new;
model  z = w;
run;
proc nlin data = reg08p03new;
parms b0 =33.82   b1 =15.74 ;
model veloc = (b0*conc)/(b1 + conc);
run;
```

Relevant Output:

Dependent Variable: z

Analysis of Variance

Source	DF	Sum of Squares	Mean Square	F Value	Pr > F
Model	1	0.40954	0.40954	652.07	<.0001
Error	20	0.01256	0.00062806		
Corrected Total	21	0.42210			

Root MSE	0.02506	R-Square	0.9702	
Dependent Mean	0.17028	Adj R-Sq	0.9688	
Coeff Var	14.71764			

Parameter Estimates

Variable	DF	Parameter Estimate	Standard Error	t Value	Pr > |t|
Intercept	1	0.02957	0.00768	3.85	0.0010
w	1	0.46549	0.01823	25.54	<.0001

Source	DF	Sum of Squares	Mean Square	F Value	Approx Pr > F
Model	2	3160.8	1580.4	7666.79	<.0001
Error	20	4.1228	0.2061		
Uncorrected Total	22	3164.9			

Parameter	Estimate	Approx Std Error	Approximate 95% Confidence Limits	
b0	27.6896	0.5247	26.5950	28.7842
b1	11.8245	0.5261	10.7271	12.9219

(a) Obtain starting values for β_0 and β_1. To do this, we can ignore the error term and note that we can transform the model into $z = \gamma_0 + \gamma_1 w$, where $z = 1/y$, $\gamma_0 = 1/\beta_0$, $\gamma_1 = \beta_1/\beta_0$, and $w = 1/x$. The initial values can be obtained from a linear regression of z on w and using $\beta_0 = 1/\gamma_0$ and $\beta_1 = \gamma_1/\gamma_0$.

A regression of z on w is performed. The estimates of the coefficients $\widehat{\gamma_0} = 0.02957$ and $\widehat{\gamma_1} = 0.46549$. Simple calculations lead to the following starting values : $1/0.02957 = 33.82$ for β_0 and $0.46549/0.02957 = 15.74$ for β_1.

(b) Using the starting values obtained in part (a), use nonlinear regression to estimate the parameters β_0 and β_1.

To estimate these parameters, we use PROC NLIN nonlinear regression option in SAS. The estimates for the coefficients are

$$\widehat{\beta_0} = 27.69$$

From the Approximate 95% Confidence Limits of β_0, we conclude that we are 95% confident that the true value of β_0 is between 26.5950 and 28.7842.

$$\widehat{\beta_1} = 11.82$$

From the Approximate 95% Confidence Limits of β_1, we conclude that we are 95% confident that the true value of β_1 is between 10.7271 and 12.9219.

Using the estimates of the coefficients, we obtain the following model

$$\widehat{\mu}_{y/x} = \frac{27.69x}{11.82 + x}.$$

Chapter 9
Indicator Variables
Solutions

3. The analysis to ascertain whether position of basketball players affects their pay over and above the effect of scoring is performed here. Data for the 1984-1985 season on pay (SAL, in thousands of dollars) and performance, as measured by scoring average (AVG, in points per game), are obtained for eight randomly selected players from each of the following position (POS): (1) scoring forward, (2) power forward, (3) center, (4) off guard, and (5) point guard.

SAS Code:
The data for this exercise is reg08p03.

```
proc format;
      value posfmt            1='scoring forward'
                              2='power forward'
                              3='center'
                              4='off guard'
                              5='point guard';
data reg09p03; **** basketball players***;
input POS   AVG   SAL;
format pos posfmt.;

cards;
 1      13.3       363
 .       .          .
 .       .          .
 .       .          .
 4       5.8        75
proc glm data = reg09p03;
class pos;
model sal = pos avg pos*avg/solution;
proc glm data = reg09p03;
class pos;
model sal = pos avg /solution;
lsmeans pos /stderr ;
run;
```

Relevant Output:

The GLM Procedure

Dependent Variable: SAL

Source	DF	Sum of Squares	Mean Square	F Value	Pr > F
Model	9	1041799.765	115755.529	3.42	0.0115
Error	19	642842.786	33833.831		
Corrected Total	28	1684642.552			

R-Square	Coeff Var	Root MSE	SAL Mean
0.618410	50.03051	183.9397	367.6552

Source	DF	Type III SS	Mean Square	F Value	Pr > F
POS	4	7433.3845	1858.3461	0.05	0.9939
AVG	1	148831.4279	148831.4279	4.40	0.0496
AVG*POS	4	31803.2359	7950.8090	0.23	0.9152

The GLM Procedure

Dependent Variable: SAL

Source	DF	Sum of Squares	Mean Square	F Value	Pr > F
Model	5	1009996.530	201999.306	6.89	0.0005
Error	23	674646.022	29332.436		
Corrected Total	28	1684642.552			

R-Square	Coeff Var	Root MSE	SAL Mean
0.599532	46.58364	171.2671	367.6552

Source	DF	Type III SS	Mean Square	F Value	Pr > F
POS	4	147636.3280	36909.0820	1.26	0.3148
AVG	1	469775.7778	469775.7778	16.02	0.0006

Parameter	Estimate	Standard Error	t Value	Pr > |t|
AVG	23.1470478	5.7839478	4.00	0.0006

Least Squares Means

POS	SAL LSMEAN	Standard Error	Pr > |t|	LSMEAN Number
center	454.714810	100.262740	0.0001	1
off guard	223.978753	76.752285	0.0077	2
point guard	415.519736	70.015761	<.0001	3
power forward	353.186774	71.151291	<.0001	4
scoring forward	401.566997	71.436262	<.0001	5

Since the data set consists of categorical variable (POS) as well as the continuous variable (AVG), analysis of covariance (ANCOVA) is appropriate. The interaction model is considered first. The p-value of the interaction between POS and AVG is 0.9152 which implies that the slopes are homogeneous. Hence, the additive model is used to analyze the data. The model is significant with an R-square of 60%. The estimated slope is 23.14

which indicate that there is a $23 (thousand) increase in salary for each one point scoring increase. It can be observed that centers get highest pay and off guards get the least.

8. A polynomial, using months as the independent variable and CDD as the dependent variable, was fit to the data in example 7.2. The data, using year as blocking variable and months as the treatment and CDD as the dependent variable, are evaluated. The appropriate polynomial in months to fit the data is found and compare with the results given in Example 7.2. The whole procedure is repeated for HDD as the dependent variable.

SAS Code:
The data for this exercise is reg07p02.

```
data powers;set reg07x02;
m2 = month*month;
m3 = month*month*month;
m4 = month*month*month*month;
m5 = month*month*month*month*month;
m6 = month*month*month*month*month*month;

proc glm data =powers;
class year;
model CDD = year month  m2 m3 m4 m5 m6/solution;
lsmeans year/stderr;
run;

proc glm data =powers;
class year;
model CDD = year month  m2 m3 m4 m5/solution;
lsmeans year/stderr;
run;

proc glm data =powers;
class year;
model HDD = year month  m2 m3 m4 m5 m6/solution;
lsmeans year/stderr;
run;

proc glm data =powers;
class year;
model HDD = year month  m2 m3 m4 m5/solution;
lsmeans year/stderr;
run;
```

Relevant Output:

CDD

The GLM Procedure

80

Dependent Variable: CDD

Source	DF	Sum of Squares	Mean Square	F Value	Pr > F
Model	10	2873649.862	287364.986	190.39	<.0001
Error	49	73958.721	1509.362		
Corrected Total	59	2947608.583			

R-Square	Coeff Var	Root MSE	CDD Mean
0.974909	16.14846	38.85050	240.5833

Source	DF	Type I SS	Mean Square	F Value	Pr > F
YEAR	4	27456.333	6864.083	4.55	0.0033
month	1	153975.603	153975.603	102.01	<.0001
m2	1	2003899.105	2003899.105	1327.65	<.0001
m3	1	271800.756	271800.756	180.08	<.0001
m4	1	313902.400	313902.400	207.97	<.0001
m5	1	98044.525	98044.525	64.96	<.0001
m6	1	4571.140	4571.140	3.03	0.0881

Parameter		Estimate	Standard Error	t Value	Pr > |t|
Intercept		-263.8833343 B	137.8558503	-1.91	0.0614
YEAR	1983	-42.7500000 B	15.8606518	-2.70	0.0096
YEAR	1984	8.1666667 B	15.8606518	0.51	0.6089
YEAR	1985	17.3333333 B	15.8606518	1.09	0.2798
YEAR	1986	10.1666667 B	15.8606518	0.64	0.5245
YEAR	1987	0.0000000 B	.	.	.
month		520.6017462	229.8194652	2.27	0.0280
m2		-343.0963229	131.9631860	-2.60	0.0123
m3		100.0239976	35.0800160	2.85	0.0064
m4		-12.7066397	4.7095419	-2.70	0.0095
m5		0.7043590	0.3097478	2.27	0.0274
m6		-0.0137908	0.0079246	-1.74	0.0881

Using year as the blocking variable, a sixth degree polynomial is fitted

$$CDD = \beta_0 + \alpha_1 I_{1983} + \alpha_2 I_{1984} + \alpha_3 I_{1985} + \alpha_4 I_{1986} + \beta_1 month + \beta_2 month^2 + \beta_3 month^3 + \beta_4 month^4 + \beta_5 month^5 + \beta_6 month^6 + \varepsilon$$

The model is significant since the p-value of F-test is less than 0.0001. The sequential (labeled Type I) sums of squares for CDD show that the sixth degree term of the polynomial is not quite significant; its p-value (0.0881) is higher than the significance level of 0.05. A fifth degree polynomial fit of the model is then run. The overall model and each of the coefficients are significant. The results are summarized below

The GLM Procedure

Dependent Variable: CDD

Source	DF	Sum of Squares	Mean Square	F Value	Pr > F
Model	9	2869078.722	318786.525	202.97	<.0001
Error	50	78529.861	1570.597		
Corrected Total	59	2947608.583			

R-Square	Coeff Var	Root MSE	CDD Mean
0.973358	16.47278	39.63076	240.5833

Source	DF	Type I SS	Mean Square	F Value	Pr > F
YEAR	4	27456.333	6864.083	4.37	0.0042
month	1	153975.603	153975.603	98.04	<.0001
m2	1	2003899.105	2003899.105	1275.88	<.0001
m3	1	271800.756	271800.756	173.06	<.0001
m4	1	313902.400	313902.400	199.86	<.0001
m5	1	98044.525	98044.525	62.42	<.0001

Parameter		Estimate	Standard Error	t Value	Pr > |t|
Intercept		-64.3924243 B	78.1123054	-0.82	0.4137
YEAR	1983	-42.7500000 B	16.1791905	-2.64	0.0110
YEAR	1984	8.1666667 B	16.1791905	0.50	0.6159
YEAR	1985	17.3333333 8	16.1791905	1.07	0.2892
YEAR	1986	10.1666667 B	16.1791905	0.63	0.5326
YEAR	1987	0.0000000 B	.	.	.
month		161.0291599	102.6461417	1.57	0.1230
m2		-126.2301959	44.2881962	-2.85	0.0063
m3		40.6167781	8.2420025	4.93	<.0001
m4		-4.5951126	0.6873828	-6.68	<.0001
m5		0.1665158	0.0210754	7.90	<.0001

We conclude that the final model has degree five

$$CDD = \beta_0 + \alpha_1 I_{1983} + \alpha_2 I_{1984} + \alpha_3 I_{1985} + \alpha_4 I_{1986} + \beta_1 month + \beta_2 month^2 + \beta_3 month^3 + \beta_4 month^4 + \beta_5 month^5 + \varepsilon$$

The parameter estimates show that the estimated model is:

$$CDD = -64.39 - 42.75I_{1983} + 8.17I_{1984} + 17.33I_{1985} + 10.17I_{1986} + 161.03month - 126.23month^2$$
$$+ 40.62month^3 - 4.60month^4 + 0.17month^5$$

Also, R-squared is equal to 0.9734 meaning that 97.34% of the variability of the cooling degree days can be explained by the variability of the month in the above fifth degree model.

HDD

The GLM Procedure

Dependent Variable: HDD

Source	DF	Sum of Squares	Mean Square	F Value	Pr > F
Model	10	2025190.330	202519.033	47.64	<.0001
Error	49	208282.520	4250.664		
Corrected Total	59	2233472.850			

R-Square	Coeff Var	Root MSE	HDD Mean
0.906745	45.44936	65.19711	143.4500

Source	DF	Type I SS	Mean Square	F Value	Pr > F
YEAR	4	25862.767	6465.692	1.52	0.2106
month	1	123041.986	123041.986	28.95	<.0001
m2	1	1764561.223	1764561.223	415.13	<.0001
m3	1	3505.478	3505.478	0.82	0.3683
m4	1	77284.575	77284.575	18.18	<.0001
m5	1	25968.654	25968.654	6.11	0.0170
m6	1	4965.648	4965.648	1.17	0.2851

Dependent Variable: HDD

Source	DF	Sum of Squares	Mean Square	F Value	Pr > F
Model	9	2020224.683	224469.409	52.63	<.0001
Error	50	213248.167	4264.963		
Corrected Total	59	2233472.850			

R-Square	Coeff Var	Root MSE	HDD Mean
0.904522	45.52575	65.30669	143.4500

Source	DF	Type I SS	Mean Square	F Value	Pr > F
YEAR	4	25862.767	6465.692	1.52	0.2118
month	1	123041.986	123041.986	28.85	<.0001
m2	1	1764561.223	1764561.223	413.73	<.0001
m3	1	3505.478	3505.478	0.82	0.3690
m4	1	77284.575	77284.575	18.12	<.0001
m5	1	25968.654	25968.654	6.09	0.0171

Parameter		Estimate	Standard Error	t Value	Pr > |t|
Intercept		674.6848485 B	128.7196038	5.24	<.0001
YEAR	1983	45.1666667 B	26.6613433	1.69	0.0965
YEAR	1984	2.5000000 B	26.6613433	0.09	0.9257
YEAR	1985	31.7500000 B	26.6613433	1.19	0.2393
YEAR	1986	-9.2500000 B	26.6613433	-0.35	0.7301
YEAR	1987	0.0000000 B	.	.	.
month		-110.8524945	169.1483899	-0.66	0.5152
m2		-69.6940980	72.9815749	-0.95	0.3442
m3		22.6563794	13.5818203	1.67	0.1015
m4		-2.3799560	1.1327234	-2.10	0.0407
m5		0.0856976	0.0347297	2.47	0.0171

Least Squares Means

YEAR	HDD LSMEAN	Standard Error	Pr > |t|	LSMEAN Number
1983	174.583333	18.852417	<.0001	1
1984	131.916667	18.852417	<.0001	2
1985	161.166667	18.852417	<.0001	3
1986	120.166667	18.852417	<.0001	4
1987	129.416667	18.852417	<.0001	5

It is found that a fifth degree polynomial should be employed for HDD. Although m3 is not significant, m4 and m5 have small p-values. Therefore, the fifth degree polynomial is appropriate.

Chapter 10
Categorical Response
Variables

1. In a study to determine the effectiveness of a new insecticide on common cockroaches, samples of 100 roaches were exposed to five levels of the insecticide. After 20 minutes the number of dead roaches was counted. Table 10.24 gives the results.

SAS Code:

```
data roaches;
input level numroach dead;
prop=dead/numroach;
cards;
5           100    15
10          100    27
15          100    35
20          100    50
30          100    69
proc logistic;
model dead/numroach=level;
output out=one p=phat;
proc sort;
      by level;
proc gplot data=one;
      symbol1 v=star c=black;
      symbol2 v=point c=black i=spline l=1;
      plot prop*level=1 phat*level=2/overlay;
run;
```

Note that the events/trials syntax was used in Proc Logistic due to the format of the data set. Variable on left of / is the number of events that occurred in number of trials given on right of slash;

Relevant Output:

The LOGISTIC Procedure

Analysis of Maximum Likelihood Estimates

Parameter	DF	Estimate	Standard Error	Wald Chi-Square	Pr > ChiSq
Intercept	1	-2.0692	0.2286	81.9590	<.0001
level	1	0.0983	0.0121	65.5659	<.0001

Odds Ratio Estimates

Effect	Point Estimate	95% Wald Confidence Limits	
level	1.103	1.077	1.130

Association of Predicted Probabilities and Observed Responses

Percent Concordant	63.5	Somers' D	0.440
Percent Discordant	19.5	Gamma	0.530
Percent Tied	17.1	Tau-a	0.210
Pairs	59584	c	0.720

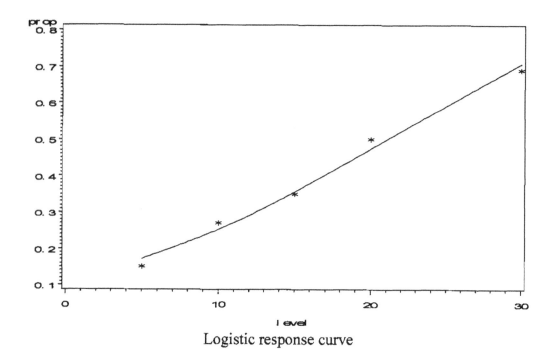

Logistic response curve

(a) Calculate the estimated logistic response curve.

The logistic regression model is estimated as follows:

$$\hat{\mu}_{y/x} = \frac{\exp(\hat{\beta}_0 + \hat{\beta}_1 x)}{1 + \exp(\hat{\beta}_0 + \hat{\beta}_1 x)}$$

$\hat{\beta}_0 = -2.0692$ and $\hat{\beta}_1 = 0.0983$

Therefore,

$$\hat{\mu}_{y/x} = \frac{\exp(-2.0692 + 0.0983x)}{1 + \exp(-2.0692 + 0.0983x)}$$

(b) Find the estimated probability of death when the concentration is 17%.

Substitute x =17 into the above model,

$$\hat{\mu}_{y/x=17} = \frac{\exp(-2.0692 + 0.0983(17))}{1 + \exp(-2.0692 + 0.0983(17))} = 0.402$$

86

Hence, the estimated probability of death for concentration level 17 % is 0.402.

(c) Find the odds ratio

The estimated slope $\hat{\beta_1} = 0.0983$ and the estimated odds ratio is calculated by $e^{\hat{\beta_1}} = 1.103$ which can be directly read from SAS output.

(d) Estimate the concentration for which 50% of the roaches treated are expected to die.

Since $\hat{\mu}_{y/x} = \dfrac{\exp(-2.0692 + 0.0983x)}{1 + \exp(-2.0692 + 0.0983x)}$

or equivalently $\ln(\dfrac{\hat{\mu}_{y/x}}{1 - \hat{\mu}_{y/x}}) = -2.0692 + 0.0983x$

Plug in $\hat{\mu}_{y/x} = 0.5$, we get

$\ln(\dfrac{0.5}{1 - 0.5}) = -2.0692 + 0.0983x \quad \Rightarrow \quad 0 = -2.0692 + 0.0983x$. Therefore, $x = 21.0498$

Hence, at the concentration level of 21.05%, we can expect 50% of the roaches treated to die.

6. The market research department for a large department store conducted a survey of credit card customers to determine if they thought that buying with a credit card was quicker than paying cash. The customers were from three different metropolitan areas. The results are given in Table 10.27. Use the hierarchical approach to loglinear modeling to determine which model best fits the data. Explain the results.

SAS Code:

```
proc catmod;
      weight n;
      model rating*city=_response_/noparm pred=freq;
        loglin rating|city @3;
proc catmod;
      weight n;
      model rating=_response_/pred=freq;
        loglin rating;
run;
```

Relevant Output:

Maximum Likelihood Analysis of Variance

Source	DF	Chi-Square	Pr > ChiSq
rating	2	54.53	<.0001
city	2	1.55	0.4597

```
rating*city              4          7.19        0.1262
```

Maximum Likelihood Analysis of Variance

Source	DF	Chi-Square	Pr > ChiSq
rating	2	55.37	<.0001

Starting with a model with both rating and city and the interaction term rating*city. That is, we fit the model

$$\mu_{ij} = \lambda_i^R + \lambda_j^C + \lambda_{ij}^{RC},$$

The PROC CATMOD was used to fit the model with the results shown above. It can be seen that the only significant variable is rating. Therefore, one variable model was used to fit the model.

$$\mu_{ij} = \lambda_i^R,$$

Notice that the likelihood ratio test for lack of fit is not significant, indicating that the model with rating only is a good fit. The p-value associated to this model is less than 0.0001.

Printed and bound by CPI Group (UK) Ltd, Croydon, CR0 4YY

03/10/2024

01040315-0018